Tropical Urban Heat Islands

Tropical Urban Heat Islands

Climate, buildings and greenery

Nyuk Hien Wong and Yu Chen

Routledge
Taylor & Francis Group

LONDON AND NEW YORK

First published 2009
by Taylor & Francis
2 Park Square, Milton Park, Abingdon, Oxfordshire OX14 4RN

Simultaneously published in the USA and Canada
by Taylor & Francis
711 Third Avenue, New York, NY 10017

Routledge is an imprint of the Taylor and Francis Group, an informa business

First issued in paperback 2015

Typeset in Sabon by
Keystroke, 28 High Street, Tettenhall, Wolverhampton

British Library Cataloguing in Publication Data
A catalogue record for this book is available from the British Library

Library of Congress Cataloging-in-Publication Data
Wong, Nyuk Hien.
  Tropical urban heat islands: climate, buildings, and greenery/Nyuk
Hien Wong. – 1st ed.
    p. cm. – (Spon research)
    Includes bibliographical references and index.
  1. Urban ecology. 2. Urban climatology. 3. Human ecology.
  4. Vegetation and climate. 5. Tropics–Climate.
  6. Buildings–Environmental engineering.
  I. Title.
  HT241.W66 2008
  307.76–dc22   2007048677

ISBN 978–0–415–41104–2 (hbk)
ISBN 978–1–138–99388–4 (pbk)
ISBN 978–0–203–93129–5 (ebk)

Contents

List of abbreviations   vii

PART I   1

1   Tropical climate   3

2   Tropical buildings   17

3   Tropical plants   35

4   Climate and buildings   51

5   Buildings and plants   68

6   Plants and climate   82

7   The plants–climate–buildings model   96

PART II   105

8   Case studies in Singapore   107

   I   UHI measurement   109

   II   Urban parks   132

   III Trees in housing developments and industrial areas   148

   IV Intensive rooftop gardens   165

   V   Extensive rooftop gardens   180

   VI Vertical landscaping   204

   VII Green experiments   227

Index   255

Abbreviations

AC	air conditioning
ASHRAE	American Society of Heating, Refrigerating and Air conditioning Engineers
BBNP	Bukit Batok Nature Park
CAM	Crassulacean Acid Metabolic
CBD	Central Business District
CBP	Changi Business Park
CFD	computational fluid dynamics
COP	coefficient of performance
CWP	Clementi Woods Park
Delta T	surface temperature difference
ECI	Equatorial Comfort Index
ET	effective temperature
ETM+	enhanced thematic mapper plus
GDP	gross domestic products
HDB	Housing and Development Board
IBP	International Business Park
IFLA	International Federation of Landscape Architects
LAI	Leaf Area Index
MRT	mean radiation temperature
MSD	Meteorological Services Division
MSS	Singapore Meteorological Service
NEA	National Environment Agency
NIR	near infra red
NParks	National Parks Board
NV	natural ventilation
OT	operative temperature
PM_{10}	particulate matter of 10 microns or less
PMV	predicted mean vote
ppm	parts per million
PRD	Park and Recreation Department
RFA	radiative forcing agents
RH	relative humidity
SD	standard deviation

SI	Singapore Index
SPM	suspended particulate matter
SPSS	Statistical Package for Social Sciences
SUHI	surface urban heat island
SWIR	short wave infra red
TC	thermal comfort
TFPC	Toronto Food Policy Council
TSI	Tropical Summer Index
UBL	urban boundary layer
UCL	urban canopy layer
UHI	urban heat island
URA	Urban Redevelopment Authority
VOC	volatile organic compounds
WHO	World Health Organisation

Part I

1 Tropical climate

1.1 Definition

The *climate* is understood as the weather averaged over a long period of time, typically 30 years (recommended by the World Meteorological Organization (WMO)). The classification of the world climates relies on some climatic parameters, such as temperature and precipitation. As one of the most widely used climate classification systems, the Köppen climate classification divides the world climates into five major types: tropical climate, dry climate, temperate climate, continental climate and polar climate. It is interesting that the different climates are selected with close reference to native vegetation, which is the best reflection of local climate. *Tropical climate*, as reflected by its name, is the hot climate typical in the tropics. The general understanding is that the area within the Tropics of Cancer and Capricorn belongs to the tropical climatic zone (see Figure 1.1). However, latitude is not the only parameter which can govern the climatic boundary. Typical tropical climate can be found beyond 23°26′. The tropical climate is of very great importance. Occupying approximately 40 per cent of the land surface of the earth, the tropics are the home to almost half of the world's population. Compared to the climate at mid-latitude, the tropical climate supports more lives and economic activities in the region. The four principal tropical areas include tropical Asia (India, Malaysia, Philippines, Indonesia, Singapore, northern Australia and so on), tropical Africa (the Congo Basin in West Africa, Mozambique, Angola and so on), tropical America (the Amazon Basin in Brazil, Argentina, Costa Rica and so on) and the tropical oceans and oceanic islands. But the climate is not uniform in the tropics. There are two broad climatic categories. One category is the warm and humid regions with excessive rainfall and considerable sunshine. It provides ideal growing conditions for luxuriant tropical plants and tropical rainforests. The other is hot and dry regions with extreme high temperature during daytime, little precipitation, low vapour pressure and low relative humidity. In these regions, deserts, semi-deserts, steppes and dry savannahs exist.

In this book, since we mainly talk about buildings and plants in dense built-up areas, we will concentrate on the hot and humid climate in which abundant tropical plants and large populations live.

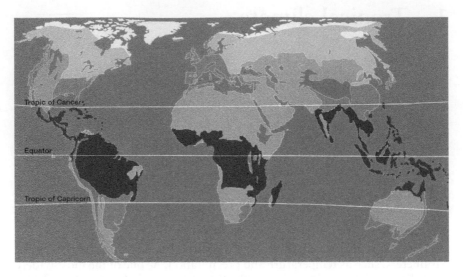

Figure 1.1 Boundary of tropical climate (hot and humid) according to the Köppen climate classification. (Picture by Yu Chen)

1.2 Negative climatic impacts

It is believed that a building in the tropics means a confrontation in terms of construction and function with extreme climatic conditions (Lauber, 2005). It is easy to summarize the extreme impacts caused by the tropical climate through its climatic parameters as follows:

1.2.1 Temperature and relative humidity (RH)

The first and most straightforward negative impact of tropical climate is its relatively high temperature and relative humidity which may cause thermal discomfort and make the tempo of life slower in the region. The reason is that the combination of high temperature and high relative humidity will reduce the rate of evaporation of moisture from the human body, especially in the locations where the lack of air movement is experienced. Due to the continuous evaporation run by the sun's heat and abundant precipitation, temperatures in the hot and humid tropics are seldom above human skin temperature during daytime. Heat loss at night, on the other hand, is limited by the abundant cloud cover. Therefore, night temperatures are maintained at a relatively high level. Such little variation of high temperature can continue throughout the year. To achieve thermal comfort, cooling strategies are always needed in built-up areas.

On the other hand, a humid microclimate condition may easily make condensation occur and lead to fungi/mould growth on building façades.

Discolouration and damage of paint and coatings may take place. Health issues are also a concern in this condition.

1.2.2 Solar radiation

In general, solar radiation received in the tropics is very high. Although the region (between 15° to 35° latitude north and south) where the greatest amount of solar radiation is received is partially beyond the tropical climate, the equatorial region between 15° latitude north and 15° latitude south still receives the second highest amount of radiation. Furthermore, the proportion of diffused radiation is very high due to high humidity and the cloud cover in the region.

Excessive solar radiation can affect the thermal conditions of buildings in two principle ways. The direct way is through openings of buildings which may heat up the interior surfaces. Absorption by building façades and eventually transferred into interior space through conduction is the indirect way which can also increase heating effects indoors. Meanwhile, solar heat absorbed by buildings during the daytime and emitted in the form of long-wave radiation at night is also a major concern in cities where the urban heat island (UHI) effect can be produced by such heat being emitted from densely placed buildings.

1.2.3 Precipitation

The amount of rainfall can be very high in hot and humid tropical areas although one or two dry seasons may be experienced. The major concern is to deal with surface runoff which may be extremely difficult in a densely built-up area. Adequate drainage from roofs and paved surfaces is of great importance. However, gutters, downpipes and other drainage systems in cities are not an economic solution for increasing water runoff. With rapid urbanization, the increase in paved surfaces becomes a heavy burden for the drainage systems in cities.

Heavy rain may also influence traffic flow and pedestrian movement. The concern when designing the openings of buildings is to avoid the negative impacts of driving rain and strong winds.

1.2.4 Wind

Wind can affect ventilation which is of great importance as a natural cooling strategy in the tropics. Weak or static air movement together with high temperature and relative humidity can easily cause thermal discomfort. It can be experienced during the inter-monsoon period or during the condition when wind-shadow is formed. Another concern is the direction of the prevailing wind and the orientation of buildings. In the tropics, eastern and western façades have been considered 'unfavorable' since high solar radiation strikes them during early morning and late evening. To minimize excessive heat gain from these orientations, it is recommended that long façades should avoid facing these directions. On the other hand, to enhance natural ventilation, long façades should be placed towards the direction of the local prevailing wind. Unfortunately, good solar orientation and good ventilation by the local prevailing wind do not always coincide. A compromise should be made to achieve a balance between the two.

1.3 Positive impacts

Because of minimal variation of diurnal and annual average air temperatures and relative humidity, architecture design in hot and humid conditions requires employment of cooling strategies to battle against heat. It means that buildings, their roofs, façades, systems and surrounding environment should all serve for the requirements of dissipating excessive heat. No heating consideration is necessary. The straightforward principles of thermal control are clearly reflected in tropical vernacular architectures (see Figure 1.2). It can be observed that the dominant characteristics of tropical buildings are openness and shading. The aim is to provide efficient ventilation and protection from the sun, rain and insects. The external walls are normally lightweight solid material (low heat capacity) with bright colour (low absorption and high reflectivity) yet ornamented beautifully. Roofs steeply sloped with an extensive overhang are also common to protect from heavy rain and hot sunshine. In some areas, the floors are elevated to keep them from the wet ground and to allow air circulation underneath. Overall, the buildings are very open in plan and construction and they are well shaded by surrounding plants.

Solar energy can be utilized in the tropics. Although there is a constant presence of cloud cover in hot and humid areas, a substantial amount of sunshine throughout the year is still experienced. This means that there is vast potential to tap into clean and renewable solar energy in the tropics, because of the following favourable conditions:

- high annual global solar radiation
- constant and uniform provision of sunshine throughout the year (no over- or under-provision occur seasonally)
- high diffused radiation which makes vertical collection possible.

Abundant rainfall, optimum temperature and sufficient sunshine make tropical plants grow extensively (see Figure 1.3). Without distinct seasons, the maintenance of plants is minimal compared to that in other climatic regions. Luxuriant plants can be found in rural areas, cities and the wilds. Greenery in a built environment can ameliorate the hot microclimate and create favourable thermal conditions for urban dwellers.

Figure 1.2 Existing 'kampong' houses in Singapore. (Photos by Yu Chen)

1.4 Thermal comfort

1.4.1 General definition

In 1966, the American Society of Heating, Refrigerating and Air conditioning Engineers (ASHRAE) introduced a definition of thermal comfort for the first time: 'Thermal comfort is the condition of mind that expresses satisfaction with the thermal environment.' In order to acquire such a sensation of comfort, human body temperature has to be maintained at a constant level internally (about 98.6 °F/37 °C). The constant temperature is achieved by continuous heat exchange between the human body and its ambient environment. In general, a thermal balance can be reached by releasing excess metabolic heat from the human body to the environment. Occasionally, the

Figure 1.3 Tropical plants. (Photos by Yu Chen)

human body can experience a short-term heat gain although it does not last long. The heat exchange is completed through four principle ways: conduction, convection, long-wave radiation exchange and evaporation (see Figure 1.4). Fanger (1970) described the steady-state heat flow per unit area per unit time by the following equation:

$$H - E_d - E_{sw} - E_{re} - L = K = R + C$$

where

H = Internal heat production in the human body
E_d = Heat loss by water vapour diffusion through the skin
E_{sw} = Heat loss by sweat evaporation
E_{re} = Sensible heat loss by evaporation
L   = Latent heat loss by respiration
K = Conduction from outer surface of the clothed body
R = Radiation loss from outer surface of the clothed body
C = Convection heat loss from outer surface.

In tropical conditions, due to high ambient air temperatures and relative humidity, some parameters, such as E_d and L, are low or negligible. Conduction loss is also small unless the body is in a reclined position. Moreover, long-wave radiation exchange is considered insignificant for heat loss in most conditions when the temperature difference between the human body and its surrounding environment is small. Thermoregulation mainly depends on convection until ambient temperature rises up to 35 °C. Evaporation is the primary solution when ambient temperature crosses the threshold of 35 °C. It is believed that increase in air movement will significantly facilitate evaporative and sensible heat losses in the tropics. Therefore, natural ventilation is of great importance in terms of achieving thermal comfort in the tropics. Besides environmental variables, wind speed, air temperature, mean radiant temperature (MRT) and so on, thermal comfort is also the function of types of clothing and activity. From a physiological point of view, on the other hand, it is also known that thermoregulation is influenced by a complicated mechanism of the human body such as temperatures of various parts of the body, skin moisture, sweat secretion, respiratory discomfort, tactile sensation fabrics and so on. However, the complex relationship between physical and physiological aspects in thermal comfort cognitive perception is still far from complete comprehension.

1.4.2 Thermal comfort in the tropics

To achieve thermal comfort in a tropical climate, or more accurately to create a comfortable environment with passive cooling strategies alone, is a real challenge. The primary reason is that the extreme climatic condition in the

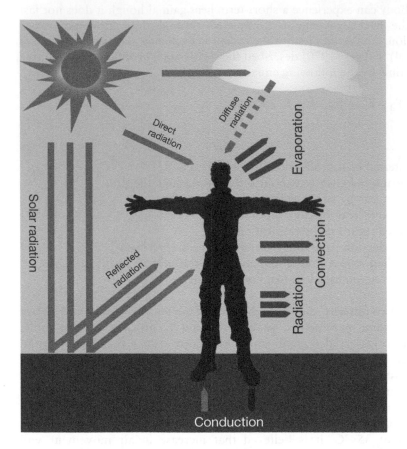

Figure 1.4 Thermal exchange between the human body and surrounding environment. (Picture by Yu Chen)

tropics is easily above the comfort zone as defined in the temperate countries (Figure 1.5). Therefore, even for a well-designed naturally ventilated building in the tropics, it is not possible to completely control the indoor thermal condition within the thermal comfort range all the time.

However, the worth of encouraging naturally ventilated design in the tropics is still there because:

- Vernacular tropical architecture shows evidence that adaptive and innovative tropical building design can still obtain fairly comfortable indoor thermal conditions.
- Tropical residents tend to accept higher temperatures physiologically and psychologically when they have certain access to environment controls, such as openable windows, light switches or curtain blinds.

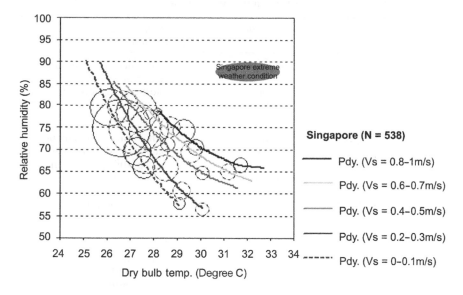

Figure 1.5 Thermal comfort chart and the extreme weather conditions in
Singapore.

Source: From Feriadi, 2003.

- Sick building syndrome can be more easily found in an air-conditioned
  building than in a naturally ventilated one.
- The occupants in a naturally ventilated building are given more options
  to take an active role in generating their most comfortable thermal
  conditions.

Four very influential factors, namely physiology, psychology, climate and
building design, have significant impacts on the thermal comfort perception
of occupants. Their influences are listed in Table 1.1.

Until now, limited thermal comfort studies have been carried out in the
tropics. The work of quantifying thermal comfort can be classified into two
major categories: field surveys and a rational approach. Table 1.2 summarizes
the studies which cover both categories.

1.4.3 Adaptive tropical perspectives

Though extensive studies have been conducted on thermal comfort in
different parts of the world, how adaptive behaviour plays a critical role in
affecting thermal comfort in the tropics remains a very interesting research
issue and much has to be done to find out more about this relationship.
It is believed that people in the tropics are able to tolerate much higher

Table 1.1 Influence of four factors on thermal comfort.

	Factors	Description	Impacts on thermal comfort
Human factors	Physiology	Metabolic rate Age Gender Weight	Body thermoregulation Acclimatization Clothing adjustment Activity (metabolic) 　adjustment
	Psychology	Behaviour Control Expectation Perception	Behavioural adjustment Background and experience Acceptability, preference Perseverance
Macro environment	Climatology	Geographical location Climate	Prevailing wind 　(monsoon wind) Diurnal humidity Outdoor temperature Solar radiation 　(sky condition) Precipitation
Micro environment	Design	Building form Openings Orientation Vegetation	Building mass (heat transfer) Porosity and shading Room dimension Window: wall ratio Building layout

temperatures and humidity than those in the temperate region because of adaptation and acclimatization. It is also believed that expectation plays a role in affecting thermal comfort as well.

Extensive thermal comfort studies in naturally ventilated environments have been conducted in Singapore to explore the impact of adaptive behaviour on thermal comfort for both indoor and semi-outdoor spaces (Song, 2007; Nyuk Hien Wong et al., 2002). Based on the ASHRAE (an indication of the thermal sensation) and Bedford (an indication of the thermal comfort) scales, this study has shown that occupants staying in naturally ventilated dwellings show higher thermal acceptability compared with what the predicted mean vote (PMV) predicted. The Bedford scale was found to be more appropriate because the scale provides a better representation of the actual thermal comfort preference and adaptive behaviour of the people in the tropics. The strong preference to vote cooler than the neutral temperature was reflected in the findings which revealed that comfortable temperatures tend to be lower than what that neutral temperature has indicated.

The studies have also shown that thermal comfort acceptability is largely influenced by non-physiological factors that go beyond the six factors which

Table 1.2 TC researches for NV buildings in tropical climates

Year	Researcher	Location	Type	Research findings
1953	(Ellis)	(Malaya) Singapore	NV	Thermal comfort (TC) survey. Physiological aspects (race, gender)
1959 1960 1961	(C. G. Webb)	Singapore	NV	TC survey, Singapore Index (SI) was derived using multiple regression analysis. Most comfortable temperature was found at 25.9 °C (SI). Equatorial Comfort Index (ECI) was later proposed
1986	(Sharma and Ali)	India (Roorkee)	NV	TC survey (18 Indian young male adults). Developed: Tropical Summer Index (TSI). Specified comfortable at 28 °C (TSI) with successive thermal sensations change at interval of 4.5 °C
1991	(De Dear, Leow and Foo)	Singapore	NV, AC	TC survey (residential, office). Discrepancy between PMV and survey thermal perception vote in NV and AC buildings. For NV: thermal neutrality (comfort) at 28.5 °C (OT)
1992	(Busch)	Thailand (Bangkok)	NV, AC	TC survey (office). For NV: the neutral temperature was found at 27.4 °C (ET) but the upper bound of acceptable level, defined by ASHRAE 55–81, was 31 °C (ET)
1996	(Mallick)	Bangladesh (Dhaka)	NV	TC criteria and building design. Comfort vote prediction
1998	(Karyono)	Indonesia (Jakarta)	NV, AC	TC survey, neutral temperature. The impacts of proper cooling on energy conservation

Table 1.2 Continued

Year	Researcher	Location	Type	Research findings
1998	(Kwok)	Hawaii	NV AC	TC survey (school). Investigated thermal neutrality, preference, acceptability. Examining TC criteria of ASHRAE 55 for tropical classrooms
1996–1999	(Nicol, Lftikhar, Arif and Najam; Nicol and Susan)	Pakistan (Five cities)	NV AC	TC survey used adaptive approach, related indoor comfort to outdoor climate The role of adaptive behaviour for TC in the office
2000	(Khedari, Nuparb, Naris and Jongjit)	Thailand (Bangkok)	NV	Proposing ventilation comfort chart Pointing out the effects of air velocity or TC
2002	(Nyuk Hien Wong et al.)	Singapore	NV	TC survey in residential buildings (255 data) Higher thermal acceptability measured by Bedford scale compared to ASHRAE scale
2003	(Nyuk Hien Wong and Khoo)	Singapore	NV	TC survey (school classrooms). Neutral temperature: 28.8 °C (OT)
2007	(Song)	Singapore	NV	Proposing Ventilation Comfort Chart for semi outdoor spaces The influence of thermal history on TC perception

have been taken into account in PMV modelling. It is believed that adaptive actions have contributed in some positive way to the higher level of thermal comfort acceptability. When occupants have the freedom to modify and adjust their environment, they always do so to compensate for the less comfortable thermal conditions. The study considered the occupant's behaviour in using the various environmental and personal controls to make themselves thermally comfortable. It revealed the occupant's tendency to modify the hot and humid living environment by creating a higher air movement (turning on fans, opening the windows). An occupant's adaptive adjustments such as drinking more water, changing their clothing and bathing more often are also favourable actions as compared with air conditioning operation.

References

Busch, J. F. (1992). A tale of two populations: thermal comfort in air-conditioned and naturally ventilated offices in Thailand. *Energy and Buildings, 18*, 235–249.

De Dear, R. J., Leow, K. G. and Foo, S. C. (1991). Thermal comfort in the humid tropics: field experiments in air conditioned and naturally ventilated buildings in Singapore. *International Journal Biometeorology, 34*, 259–265.

Ellis, F. P. (1953). Thermal comfort in warm humid atmosphere observations on groups and individuals in Singapore. *Journal of Hygiene 50*, 386–404.

Fanger, P. O. (1970). *Thermal Comfort: Analysis and Applications in Environmental Engineering.* New York: McGraw-Hill.

Feriadi, H. (2003). *Thermal Comfort for Naturally Ventilated Residential Buildings in the Tropical Climate.* Singapore: National University of Singapore.

Karyono, T. H. (1998). Report on thermal comfort and building energy studies in Jakarta-Indonesia. *Building and Environment, 35*, 77–90.

Khedari, J., Nuparb, Y., Naris, P. and Jongjit, H. (2000). Thailand ventilation comfort chart. *Energy and Buildings, 32*, 245–249.

Kwok, A. G. (1998). Thermal comfort in tropical classrooms. *ASHRAE Transactions, 104*(1B), 1031–1047.

Lauber, W. (2005). *Tropical Architecture: Sustainable and Humane Building in Africa, Latin America and South-East Asia.* Munich: Prestel.

Mallick, F. H. (1996). Thermal comfort and building design in the tropical climates. *Energy and Buildings, 23*, 161–167.

Nicol, J. F., Lftikhar, A. R., Arif, A. and Najam, J. G. (1999). Climatic variations in comfortable temperatures: the Pakistan Projects. *Energy and Buildings, 30*, 261–279.

Nicol, J. F. and Susan, R. (1996). Pioneering new indoor temperature standards: the Pakistan Project. *Energy and Buildings, 23*, 168–174.

Sharma, M. R. and Ali, S. (1986). Tropical summer index – a study of thermal comfort of Indian subjects. *Building and Environment, 21*, 11–24.

Song, J. F. (2007). Thermal comfort in semi-outdoor spaces in hot and humid tropics (unpublished PhD thesis). National University of Singapore, Singapore.

Webb, C. G. (1959). An analysis of some observations on thermal comfort in an equatorial climate. *British Journal of Industrial Medicine, 16*, 297–310.

Webb, C. G. (1960). Thermal discomfort in an equatorial climate. *Journal of the Institution of Heating and Ventilating Engineers, 16*, 297–310.

Webb, C. G. (1961a). A comfort graph for life in the tropics. *The New Scientist, 8,* 1643–1645.

Webb, C. G. (1961b). The diurnal variation of warmth and discomfort in some buildings in Singapore. *Annual Occupancy Hygine, 3,* 205–218.

Wong, N. H., Feriadi, H., Lim, P. Y., Tham, K. W., Sekhar, C. and Cheong, K. W. (2002). Thermal comfort evaluation of naturally ventilated public housing in Singapore. *Building and Environment, 37,* 1267–1277.

Wong, N. H. and Khoo, S. S. (2003). Thermal comfort in classrooms in the tropics. *Energy and Buildings, 35*(4), 337–351.

2   Tropical buildings

2.1 Urban context

Urbanization is the growth in the proportion of the population living in urban areas. The world has experienced unprecedented urban growth in the last and current centuries. In 1800, only 3 per cent of the world's population lived in urban areas. The world population began to increase substantially after 1900. The percentage of urban population increased 14 per cent and 47 per cent in 1900 and 2000 respectively. For the first time in history, more than half of the world population is living in urban areas in 2008 (Laski and Schellekens, 2007). It is also estimated that by 2030, up to 5 billion people will live in towns and cities.

The truth is that almost half of the world population lives in the tropics. Figure 2.1 shows the rate of urbanization in the tropical areas. According to Gupta (2002), the urban population of the developing countries increased rapidly from 286 million to 1,515 million between 1950 and 1990 and the figure will reach up to 4 billion by 2025, with almost all developing countries within the tropics and subtropics. Therefore, significant attention should be paid to urbanization in the tropics.

The pace of urbanization in the tropics over the past 60 years has been much faster than that of the developed countries, which took about 200 years in the moderate region. From colonial times to independence, especially in the past 50 years, many countries in the tropics have made significant investment to build more buildings and improve the conditions of infrastructure and architecture in their cities. However, architectural and structural design have paid very little attention to the local extreme climatic conditions in the region. Bay (2006: 46) believed that:

> Unfortunately, much of the work that passes for architecture in the tropics today are unadulterated transplants from temperate countries, particularly the US – justified in the name of International Style.

It is regrettable that the architectural style generated in the moderate environment has been transplanted with little recognition of tropical climate

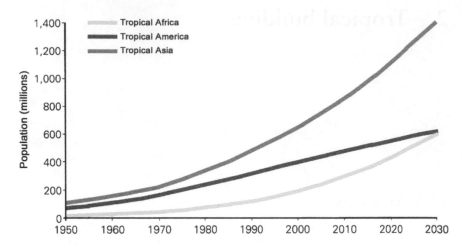

Figure 2.1 Rate of urbanization in the tropics.

Source: Adapted from Emmanuel, 2005: 6.

Notes: Tropical Africa includes Western, Middle and Eastern Africa; Tropical Asia includes South and Southeast Asia; Tropical America includes Central and tropical South America plus the Caribbean. Due to less than 2 million urban population in 2000, Tropical Oceania is not included in the figure.

conditions. Together with the idea of employing high-rise and high-density strategies to solve the issue of overcrowding, some big tropical cities, such as Singapore, Kuala Lumpur, Bangkok, Jakarta and São Paulo, pose many environmental and social issues.

2.1.1 Pollution

Air or water pollution can be found in every big city. The uniqueness of tropical cities is probably the relatively high ambient air temperature and relative humidity throughout the year. It is believed that atmospheric pollution can be aggravated by the accumulation of smog that is related to the combination of the higher temperature and the presence of air pollutants. As the macro level wind is usually weak in tropical cities, pollution dispersal can be especially difficult. It has been found that the level of suspended particulate matter (SPM) often exceeded the World Health Organization (WHO) guidelines by a factor of two and the concentration level of SPM often exceeded the guidelines by a factor of four (Emmanuel, 2005: 7–8). Among radiative forcing agents (RFA), ozone is the major pollution in tropical cities. With enough ultraviolet sunlight and moisture (commonly found in tropical environments), ozone can be easily formed through a photochemical reaction of nitrogen oxides and volatile organic compounds (VOC). The discharge of storm water in tropical cities, due to high temperature and the pollutants carried, can increase biological oxygen demand which is a threat to aquatic life.

2.1.2 Climate change

The change of urban climate, especially micro-climate, is definitely associated with the rapid urbanization. One of the representative examples is the emergence of the urban heat island (UHI) effect in tropical cities. Unlike in temperate cities, the UHI effect can bring negative impacts all the time in the tropical condition. Higher temperatures in urban areas means hazards of thermal discomfort, air pollution and even water pollution. Due to the expansion of cities' territory, the magnitude and intensity of the UHI effect in the tropics attracts more serious attention. The UHI effect will be further discussed in Chapter 8 and a survey of the UHI effect in Singapore will be shown in Case I.

2.1.3 Excessive cooling energy consumption

It is believed that air conditioning systems are the main consumer of energy in a tropical environment. Typically, for a commercial building in the tropics, the air conditioning system alone will consume more than 50 per cent of the total energy consumption in a building (see Figure 2.2).

As society becomes more affluent, the use of air conditioning is becoming pervasive and this has resulted in a drastic increase in the demand for electricity consumption. Figure 2.3 shows that there is an exponential

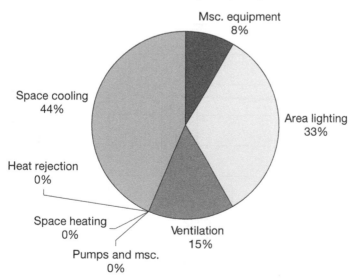

Figure 2.2 Annual energy consumption by enduse in the tropics.

increase in the electricity demand, far greater than the increase in GDP for Singapore. The electricity consumption of Singapore has increased from 25,858 GWh in 1998 to 41,017 GWh in 2007. This represents an annual increase of 5.3 per cent.

Facing the issues raised by contemporary buildings in tropical cities, it is necessary to rethink city planning in the context of the tropics. Givoni (1994) believed that many features of the physical structure of a tropical city can have great impact on the urban climate. It is possible to improve the urban climate through proper urban planning and design. Important design elements include location of the cities in a region, layout/orientation of a street's network, density of a built-up area, types of buildings and green areas. Tay (2001) mentioned that 'the mega-cities of the tropics are not the result of natural evolution. They have come about within the last 200 years as a result of colonial intervention.' He also believed that the tropical concept cannot be reflected by one single building but by a group of buildings which are properly designed in the urban tropics. He therefore raised a new conceptual framework – the tropical city concept which aims to positively make use of the sun, rain, wind and vegetation to imaginatively produce a conducive and efficient living environment. Emmanuel (2005) recommended that the common spaces between buildings in the urban tropics should be designed with sensitivity to the tropical climate. Two major targets of a successful tropical urban design, which should be carried out not only at the building scale but also the urban one, are reducing solar heat gain and promoting natural ventilation.

Every tropical specialist mentioned above highlighted the importance of plants in the urban tropics. Preserving and replicating greenery can bring

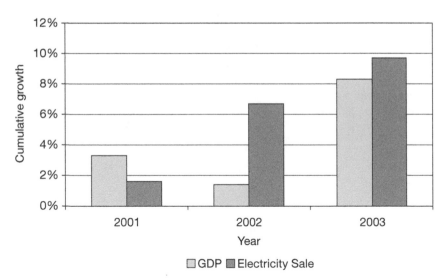

Figure 2.3 Trends of cumulative growth in energy demand with GDP.

Source: *Singapore Yearbook of Statistics* published by Singapore Department of Statistics

many related benefits to the cities. However, it can be a mixed bag since plants can provide not only welcome shading but also annoying wind blockage and humidity. Strategically introducing plants into a built environment is of great importance in the context of city planning.

2.2 Vernacular buildings in the tropics

'Vernacular buildings have evolved over time to make the best use of local materials and conditions to provide adequate, and often luxurious, shelter for populations inhabiting even the most extreme climates of the world' (Roaf et al., 2005: 33). Without any doubt, valuable lessons can be learned from the vernacular buildings in the tropics (see Figure 2.4). The point is not only to appreciate their colourful architectural features but also to digest the way they respond to the local climate.

The Kalae house, Bamboo house, Kampong house and Torajan house are four typical traditional houses selected from Thailand, the Philippines, Malaysia and Indonesia respectively. They have their own unique architectural features. For instance, the Kalae house got its name from the V-shaped wooden decoration, representing the horns of buffalo, erected at the gable end peaks (Sthapitanonda and Mertens, 2005). The Bamboo house is simply made of bamboo. Bamboo walls and floors make air circulation from outside to inside easy through the slats. The Malay Kampong house is special in its layout (Vlatseas, 1990). It usually constitutes at least two parts: the main space (Rumah Ibu) and kitchen annex (Rumah Dapur). Such an arrangement can reduce the damage caused by a fire in the kitchen. A raised veranda (Serambi) is also commonly seen in the Malay Kampong house. The Torajan house (also known as Tongkonan) is the traditional house of the Torajan people who live in Sulawesi, Indonesia (Kis-Jovak et al., 1988). The Torajan house is famous for its boat-shaped saddleback roof and gables which are dramatically upswept.

Although they are from different countries, these traditional houses are facing the same climatic conditions – high solar radiation, high temperature and relative humidity, small diurnal temperature variation, heavy precipitation and so on. Therefore, they exhibit some similar characteristics:

- they are commonly raised on stilts or piles
- they have steep roofs
- they are well-oriented to avoid excessive solar gain
- they are built near to plants.

The tropical vernacular buildings can reflect unique climate-oriented design in the ways listed below.

2.2.1 By using local materials

Timber or bamboo walls and floors are commonly seen in these traditional houses, as the rainforest provides sufficient plant materials. The obvious

Figure 2.4 Traditional houses in the hot and humid climate. Left upper: Kalae house, Thailand; Right upper: Bamboo house, Philippines; Left lower: Kampong house, Malaysia; Right lower: Torajan house, Indonesia. (Pictures by Yu Chen)

advantage is that these lightweight structures have low thermal capacity which reduces solar heat accumulation during the daytime and long-wave heat release at night. Raw plants are used as roofing materials as well. Thatch is commonly found to be a durable and insulated roofing material in the region. It has low thermal conductance yet high insulation value since a large amount of air is trapped inside the thick roof. Convective heat loss can be increased when the roof has a large uneven surface, which is commonly observed in the traditional house as a large 'umbrella'. Furthermore, the thatch roof can also trap water from either rain or morning dew and evaporate it though solar heat during the daytime. The 'breathing' process naturally cools the building and avoids the condensation problem.

2.2.2 By layout/orientation design

Most of the vernacular buildings are properly arranged with their long axis north–south oriented. The aim is to prevent excessive heat gain at the eastern and western orientations. The layouts of these tropical houses are mostly open and permeable which reflect the desire to introduce ventilation throughout the indoor space.

2.2.3 By structure design

Roofs are the dominant element in these buildings. With steep slopes and large overhangs, the roofs can simultaneously provide adequate sun shading and shed rain rapidly and silently. Meanwhile, the deep roofs can also create a 'stack effect' inside the buildings, which encourages air circulation indoors. It can be observed that most of these houses are built on stilts. The aims are very straightforward: raising the structure above the wet ground, allowing air circulation underneath, exposing it to wind at higher levels and protecting it from frequent floods, soaking and wild animals. Pivoted windows can be opened outwards. They can serve as sun shades as well as wind baffles.

2.2.4 By being close to nature

The traditional houses merge into the landscape, which easily awakens the sense of nature. They are purposely built to be surrounded with luxuriant plants. Since the houses are low-rise structures, trees can cast efficient shading on them and create a tolerable micro-climate in residential areas. Again, plants have been considered as a significant element in these tropical vernacular houses.

These simple yet effective climate-sensitive designs are the result of a long-term evolution of the venerable houses in the tropical climate. They provide a model for the design of succeeding tropical architecture. Unfortunately, they lack the sophistication of modern life. Today, these traditional buildings have given way to modern concrete and glass buildings which can fulfil the requirements of rapid urbanization and increased population in cities.

2.3 Contemporary buildings

Most tropical countries have experienced a similar development from ancient times as quiet agricultural lands, to the colonial era as countries at the cross-roads of the east–west trade, to the eventual period of independence and development. Rapid growth mostly took place when the countries began on their roads to political independence and economic development. It is well known that 'the hot-humid tropics is experiencing both unprecedented and unique urban growth' (Emmanuel, 2005: 5). The rapid urbanization has been reflected by the tremendous expansion of tropical cities. Infrastructures and buildings are being demolished, built and rebuilt everywhere. However, not all buildings take into consideration the extreme climatic conditions in the tropical world. Uncritical acceptance of Western building forms and overconfidence in modern techniques are probably the main reasons for duplicating identical building forms in the region. The consequence is high running costs or unfavourable indoor thermal conditions.

Singapore, as a highly developed tropical city with varied buildings reflecting its ethnic build-up, is a good example to illustrate tropical modern buildings (see Figure 2.7). Bay (2006) briefly described some representative tropical architecture where local architects had tried to interpret architectural designs in the context of tropical conditions.

> Since 1819, the British started to build colonial bungalows with European constructional methods and materials heuristically adapted to the local climate in Singapore. . . . The streets in Singapore were characterised by *shophouse*. . . . The shophouses were ventilated via internal courtyards and with "jacked-roofs" at the ridge of the roofs. . . . The colonial way of building continued up to the end of 1950's.
>
> (Bay, 2006: 23)

Many of these colonial type of buildings were destroyed during the post-war decades. Some remaining buildings can be found in Chinatown and some are scattered throughout the downtown area (see Figure 2.5). Similar to the vernacular houses, it is believed that the colonial architecture is still highly instructive since the local climate was carefully considered in these buildings.

> In 1937, inspired by contemporary European glass architecture, the Public Works Department designed and built the Singapore Kallang Airport Terminal Building as one of the first new generation of buildings. Its way of adapting to the tropical climate was by providing large pronounced horizontal fins over the large areas of horizontal glass façades, offering great expanse of shelter from the sun and rain. . . . Ng

Figure 2.5 Shophouse in Chinatown, Singapore. (Photos by Yu Chen)

Keng Siang ... designed the tallest high-rise building in Singapore's business district in 1954 ... by adapting the façade design to the local climate with articulated horizontal shading strips.

(Bay, 2006: 24)

The shift away from colonial tropical architecture and its replacement with a modern tropical architecture became more widespread during the post-colonial period. ... the dawn of Singapore's nation building saw several major modern architectural endeavours. These include the Singapore Conference Hall and Trade Union House (a very important public building), the Malaysia Singapore Airways Building (the prototype non-colonial, high-rise commercial building in Singapore's main business street), the People's Park Complex (the first high-rise complex with high-density mixed-development of housing and shopping complex) and the Who Hup complex (the experimentation of urban architecture for the new tropical Asian city).

(Bay, 2006: 24)

All these buildings illustrate the attempt to create modern tropical buildings with respect to the local climate (see Figure 2.6). Sun shading devices and natural ventilation were integrated into the designs.

Figure 2.6 Contemporary Singapore buildings reflecting the tropical design concepts. Upper left: Singapore Conference Hall and Trade Union House; upper right: Singapore Kallang Airport Terminal Building; lower left: Who Hup Complex; lower right: People's Park Complex. (Photos by Yu Chen)

Climatically appropriate architecture can also be found in the predominant public housing style, the Housing and Development Board (HDB) flats in which 3 million out of a 3.6 million population are currently accommodated (HDB Annual Report 2005/06). With the objective of constructing low-cost public housing, the Housing and Development Board began to build public housing from 1961. The primary concern in HDB block layout design was to minimize solar penetration in the tropical climate. The long axes of the blocks are therefore mostly north–south oriented. Void deck on the ground floor can also be found in some HDB flats, especially the early batches (see Figure 2.8). This feature greatly enhances natural ventilation through directing the cool breeze into the inner part of the community. A study has been conducted in Singapore to examine the impact of void decks on the pedestrian wind velocity. Figure 2.9 shows that with the presence of full-height void decks, the wind velocity can be increased by as much as 100 per cent. The presence of void decks can also enhance the natural ventilation performance of the units immediately above the void decks.

To protect from strong solar radiation on façades, long corridors/verandas and sun-shading devices are also commonly seen in the HDB houses (see Figure 2.10).

Figure 2.7 Skyline of Singapore. (Picture by Yu Chen)

Figure 2.8 HDB void decks. (Photos by Yu Chen)

It is important to notice that more climatic concerns have been shown in old HDB flats than in new ones. The possible reason is that current designers/architects rely too much on technical-based concepts, such as an air conditioning system, but ignore the use of natural principles of cooling.

One trend has been the pervasive use of full-height glazing for residential buildings without the proper provision of shading (see Figure 2.11). A simulation study has been conducted to examine the implications of the provision of shading on thermal comfort and cooling energy. Figure 2.12 shows that with the provision of shading/balcony, the difference between mean radiant temperature and air temperature can be drastically reduced from 8°C to less than 2°C. This will enhance the thermal comfort extensively. Figure 2.13 also shows the provision of shading on cooling energy consumption. For example, with the provision of 1.5-metre shading, the

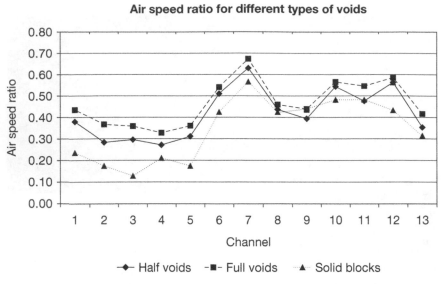

Figure 2.9 The impact of void decks on the pedestrian wind velocity in Singapore.

Figure 2.10 Sun-shading devices constructed in a local HDB block.

Figure 2.11 A condominium with full-height glazing design (residents have to use curtains, shutters and so on to prevent themselves from strong incident solar radiation. (Photo by Yu Chen)

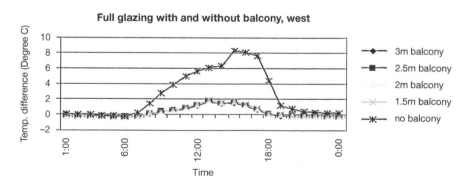

Figure 2.12 Temperature difference between mean radiant temperature and air temperature with and without balcony at western orientation.

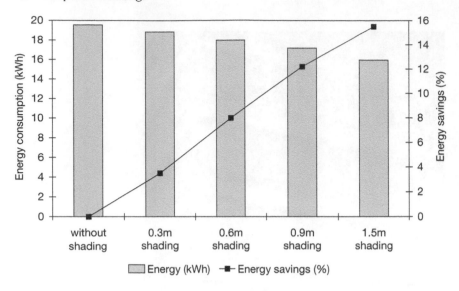

Figure 2.13 Cooling energy savings by sun shading.

saving on cooling energy consumption can amount to 15.5 per cent compared to no provision of shading.

Similar problems have also been found recently in commercial buildings and private condominiums in Singapore (see Figure 2.14). 'The use of numerous new techniques has allowed architecture to escape the restraints imposed by places and materials, indeed to liberate itself from them entirely' (Lauber, 2005: 29). Although diverse designs have replaced the earlier rigid ones, some modern buildings begin to show very little response in terms of architecture and construction to the extreme climatic conditions. The reasons for building climatically 'sick' buildings can be summarized as follows:

- wrongly oriented
- inappropriate building forms (including façades inclined towards the sun, large areas of glazing and inappropriate building materials)
- lack of or wrongly placed sun-shading devices. For example, the use of shading devices with extensive vertical fins in west-facing façades may not be effective as the angle of the sun in the tropics tends to be high most of the time
- lack of cross ventilation.

Regardless of its aesthetic appearance, contemporary tropical architecture has shown a mix of climatically appropriate and inappropriate forms. The former are represented by the colonial buildings and some 'old' architecture. It is understandable since technology-based concepts were not mature when

Figure 2.14 Some problematic buildings, such as inappropriate building forms and
wrongly placed sun-shading devices. (Photos by Yu Chen)

the 'old' buildings were built. In order to achieve thermal comfort, they had to show respect for the local climate in a passive manner. The latter are mostly found in relatively 'new' buildings which are completely air-conditioned and cut off all the natural relationship to the climate. The price is a waste of valuable primary energy and the rise of many related side-effects, which will influence not only the building itself, but also the whole urban environment. This does not mean that the use of technology-based concepts in buildings is wrong. The point is how to achieve sustainable performance with the combination of techniques and natural principles.

2.4 Design strategies

In hot and humid conditions, two fundamental objectives should be strictly observed. One is to minimize solar heat gain through employing proper shading or materials and the other is to maximize evaporative cooling through introducing natural ventilation. The two objectives apply to design carried out at both urban scale and individual buildings.

Some general urban design strategies should be followed:

- The urban space between buildings should be dimensionally proper for encouraging air movement. This is closely related to the density of urban areas, the dimension of urban canyons, openness, the orientation and the width of streets.
- Neighbourhoods should be well oriented to avoid bad solar orientations but welcome prevailing wind. The priority should be given to achieving good cross-ventilation when there is a conflict between solar and wind orientation.
- Self-shading and shading from nearby buildings should be well utilized within sites. According to Emmanuel's 'shadow umbrella' strategies (Emmanuel, 2005: 114), tall buildings are better arranged along the north and east while low buildings or plants are arranged along the west. Courtyards in the centre can be left open.
- Light colour (low absorptivity and high reflectivity) façades should be encouraged in a built environment (mitigating the UHI effect).
- Vegetation should be introduced extensively yet carefully in urban areas. This serves two major functions: providing shading at pedestrian level and building façades or roofs. However, blockage of air movement should be avoided.

Some general building design strategies should be followed:

- Individual buildings should employ open layout.
- The long axis of the building should be placed along a north–south orientation.

- The long axis of the building should be orientated approximately perpendicular to the local prevailing wind. If it conflicts with solar orientation, the priority should be given to achieving good natural ventilation.
- Large openings are welcomed at façades to encourage air movement. But light and reflective sun-shading devices must be applied to protect the openings from solar radiation and unwanted light without blocking air movement. Windows should be openable to allow natural ventilation even for air-conditioned rooms. Openings are best placed on opposing walls to facilitate cross-ventilation.
- The building envelope should employ lightweight construction with small heat capacity to reduce heat storage maximally. A light colour envelope is encouraged to reduce the absorption and thus the transfer of heat indoors.
- Arched, domed and pitched roofs are better than flat ones in terms of heat gain/loss and rainwater removal. Large roof overhangs are encouraged to protect walls and openings from solar heat and rain.
- Raising the first floor off the ground is encouraged to allow cool air circulation beneath.
- Plants should be strategically introduced into buildings, in their roofs, façades and surroundings, to provide their unique thermal benefits.

In order to design a climatically appropriate city and individual architecture in the tropical climate, all the general strategies should be considered carefully. In this book, the focus has been placed on the area of strategically introducing plants into a built environment although other strategies are of similar importance.

References

Bay Joo Hwa, O. B. L. (ed.) (2006). *Tropical Sustainable Architecture – Social and Environmental Dimensions*. Oxford: Architectural/Elsevier.

Emmanuel, M. R. (2005). *An Urban Approach to Climate Sensitive Design: Strategies for the Tropics*. London and New York: Spon Press.

Givoni, B. (1994). Urban design for hot, humid regions. *Renewable Energy*, 5(5–8), 1047–1053.

Gupta, A. (2002). Geoindicators for tropical urbanization. *Journal of Environmental Geology*, 42(7), 736–742.

Joo-Hwa, B. (2001). *Cognitive Biases in Design: The Case of Tropical Architecture*. The Netherlands: Technische Universiteit Delft, Delft.

Kis-Jovak, J. I., Nooy-Palm, H., Schefold, R. and Schulz-Dornburg, U. (1988). *Banua Toraja: Changing Patterns in Architecture and Symbolism among the Sa'dan Toraja, Sulawesi, Indonesia*. The Netherlands: Royal Tropical Institute.

Laski, L. and Schellekens, S. (2007). Growing up urban. In A. Marshall and A. Singer (eds), *The State of World Population 2007 Youth Supplement*. United Nations Population Fund (UNFPA).

Lauber, W. (2005). *Tropical Architecture: Sustainable and Humane Building in Africa, Latin America and South-East Asia*. Munich: Prestel.

Roaf, S., Chrichton, D. and Nicol, F. (2005). *Adapting Buildings and Cities for Climate Change: A 21st Century Survival Guide*. Oxford and Burlington, MA: Architectural Press.

Sthapitanonda, N. and Mertens, B. (2005). *Architecture of Thailand: A Guide to Traditional and Contemporary Forms*. Singapore: Editions Didier Millet.

Tay, K. S. (2001). Rethinking the city in the tropics: the tropical city concept. In A. Tzonis, L. Lefaivre and B. Stagno (eds), *Tropical Architecture: Critical Regionalism in the Age of Globalization* (pp. 266–306). Chichester: Wiley-Academy.

Vlatseas, S. (1990). *A History of Malaysian Architecture*. Singapore: Longman Singapore.

3   Tropical plants

3.1 Tropical rainforest

The natural formation which once fully covered the hot and humid tropical areas is the tropical rainforest which is generally found near the equator. Its name (derived from the German *tropische Regenwald*) reflects a very important climatic parameter – rainfall. Basically tropical rainforest can only be found in a hot and humid location with abundant rainfall. Yamada (1997: 1) believed that 'this is a world comprising a multi-layered society of plants reaching from the ground to the treetops 70 metres above, a world that hosts a stable society in the midst of the most complex diversity on earth'. Due to human activities and natural disasters, the remaining tropical rainforest covers only 7 per cent of the earth's dry land, distributed in three major areas (see Figure 3.1). The significant impacts of the tropical rainforest, however, cannot be ignored:

- It is the habitat which supports more than half of all the world's plant and animal species including large numbers of unknown species (Dirzo and Raven, 2003).
- It can purify the air as the earth's green lungs.
- It has a significant role in global carbon and energy cycles (IPCC, 2002).
- It can deter the greenhouse effect through storing carbon dioxide in roots, stems, branches and leaves.
- It can provide humans with food, fuel wood, building materials and medicines.

Forest cleaned and replaced by an alternative land cover is called 'deforestation', which is caused by human socio-economic activities (such as commercial logging, farming and ranching), fire and drought. For instance, within one year from 1997 to 1998, 5 million hectares of forest were burned for agriculture and tree plantation purposes in Sumatra and Kalimantan, Indonesia (Sodhi and Brook, 2006). The fire was facilitated by dry conditions and resulted in a bad haze in Southeast Asia. Unfortunately, people do not yet realize the adverse consequences and the loss of the tropical rainforest still proceeds at an amazing pace. The highest rates of tropical deforestation

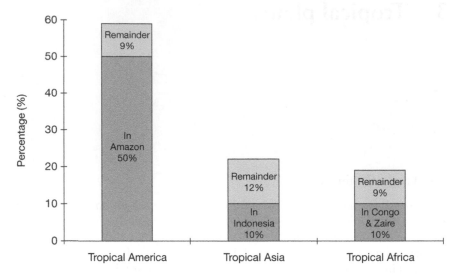

Figure 3.1 Tropical rainforest in the world.
Source: Glantz and Krenz, 1992: 588.

recorded so far were observed during the 1980s and 1990s (Ramankutty and Foley, 1999). In tropical Asia, deforestation rates increased from 1.8×10^6 ha per year during the 1980s to 2.6×10^6 ha per year during the 1990s (Hansen and DeFries, 2004). The rates in the Amazon area, where the largest tropical rainforest exists, rapidly increased from 1.7×10^6 ha per year during the 1990s to 2.4×10^6 ha per year in 2003 (Fearnside and Barbosa, 2004). Figure 3.2 shows the rate of deforestation in pantropical areas observed during the period from 1990 to 1997 through remote sensing technology.

The deforestation has the following possible negative impacts according to McGregor and Nieuwolt (1998):

• climate change
• displacement of indigenous cultures
• the loss of biodiversity
• the loss of soil fertility
• the degradation of the quality of water resources.

Furthermore, the destruction of the current rainforests may release greenhouse gas – carbon dioxide – which is stored in plants and worsen the already severe global warming issue. Careful planning, sensitive harvesting and appropriate agriculture are the only ways to achieve a sustainable future for the tropical rainforest.

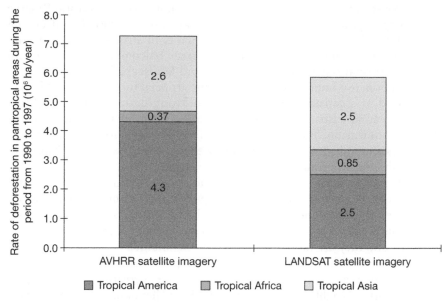

Figure 3.2 The rate of deforestation in pantropical areas including Africa, America and Asia.

Source: Based on data from Wright, 2005.
(AVHRR data derived from Advanced Very High Resolution Radiometer 8 km resolution satellite imagery for all lands between the tropics; LANDSAT data derived from LandSat 30 m resolution satellite imagery for the 'evergreen and seasonal forest of the tropical humid bioclimatic zone' for all continents plus the 'dry biome of continental Southeast Asia' but exclusive of Mexico and the Atlantic coastal forest of Brazil.)

3.2 Urban green

Simultaneously with the rapid deforestation, rapid urbanization is taking place in the tropical area. Some invaluable forests have to give way to the expansion of the built environment to accommodate increasing human population. But vegetation always accompanies the growth of cities in different formations. It is difficult to imagine a city without any plants even in the centre of the densest 'concrete jungle'. The reason for keeping green colour in a city is not only because plants are the aborigines which should be preserved but also because their broader benefits cannot be produced by any other life-form. Plants provide economic as well as environmental and aesthetic advantages to the urban environment (this will be discussed in-depth in Chapter 5).

In a city, the formations of native plants can be roughly divided into two major categories: natural and artificial (see Figure 3.3). Natural formation indicates those aboriginal plants which are preserved in a built environment. Normally wildlife, flora, fauna or geological features will be protected together within an area called a natural reserve. It is rare in a built environment due to the constraint of space. One example is the Bukit Timah Natural Reserve in Singapore City. The majority of the plant life in cities should belong to artificial formations. Parks, gardens, plants in streets, courtyards and open spaces, green roofs, green walls and green balconies or terraces are all artificial formations which are planned and landscaped in the process of urbanization. It is important to note that the artificial formation should never be viewed as a satisfactory alternative to losing nature which should be preserved at all costs. This artificial formation is simply the compromise to rapid urbanization. As a precious resource, artificial formations are the windows and links from which urban dwellers can access Mother Nature in the harsh built environment.

Artificial formations can be further divided into two groups. One is landscaping on the ground which fills in the public areas in urban environment. According to their dimensions, they can be simply classified as city parks, neighbourhood parks and other green areas. City parks are features with a big area and they are normally owned and maintained by local government (see Figure 3.4). The purpose of building such parks can vary from preserving a unique landscape, to providing active or passive recreation, to shaping the image of cities. Neighbourhood parks, usually smaller in size, are more flexible green areas in-between developments (see Figure 3.5). As community assets, they are normally built to beautify the harsh built environment and to provide immediate access to nature for residents. Other green areas (see Figure 3.6) are indeed smaller areas of landscape in-between urban spaces in different forms such as plants in playgrounds and courtyards and trees along roadsides. Except for aesthetics considerations, they are commonly built to protect specific locations (playgrounds, roads, etc.) from excessive solar exposure, noise, wind and so on.

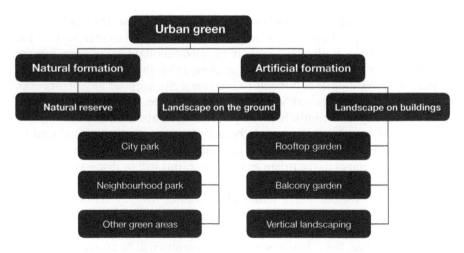

Figure 3.3 Formations of urban green and greening considerations.

Figure 3.4 Bukit Batok Natural Reserve, Singapore. (Photos by Yu Chen)

Figure 3.5 Neighbourhood parks. (Photos by Yu Chen)

The other group under artificial formations is landscaping on buildings. This includes rooftop gardens, balcony/terrace gardens and vertical landscaping. Two major types of rooftop garden exist at the moment: intensive and extensive. Intensive rooftop gardens, on which various species of plants can grow, require relatively thick growing media and extensive maintenance (see Figure 3.7). They are normally accessible as a recreation space for residents. Hence, they usually incorporate areas of paving, seating and other architectural features. Extensive green roofs, on which only low maintenance and self-generative plants (turfing/grass is common) can be planted, feature lightweight growing media and low maintenance (see Figure 3.8). Without access, they are mainly designed for aesthetic and environmental purposes rather than entertaining residents. Besides on flat roofs, in some European countries, extensive systems can also be found on pitched roofs. A balcony/terrace garden is a smaller garden incorporated in balconies or small terraces (see Figure 3.9). The method of growing plants can be implemented through garden pots or a specially designed structure.

Figure 3.6 Other green areas. (Photos by Yu Chen)

The balcony/terrace garden serves as a natural link between interior and exterior environments. Such a garden is tended by individual residents in residential buildings or by a management agency in commercial buildings. The maintenance work is normally not very heavy.

Compared with green roofs, vertically placed greenery can cover more exposed hard surfaces in a densely built-up environment where high-rise buildings are the predominant building style. Ken Yeang (1998: 103) believed that

> This will significantly contribute towards the greening of the environment if a skyscraper has a plant ratio of one to seven, then the façade area is equivalent to almost three times the site area. So if you cover, let's say, even two-thirds of the façade you have already contributed towards doubling the extent of vegetation on the site. So in fact a skyscraper can become green. And if you green it, you're actually increasing the organic mass on the site.

Plants introduced vertically into buildings is not a new concept. However, introducing plants into building façades is strategically still a challenge due to the lack of R & D in the area.

Figure 3.7 Intensive rooftop gardens. (Photos by Yu Chen)

According to the species of plants, types of growing media and construction methods, vertical landscaping can be simply divided into three fundamental types: the wall-climbing type, the hanging-down type and the module type (see Table 3.1). The wall-climbing type is a very common and traditional vertical landscaping method (see Figure 3.10). Climbing plants can cover the walls of buildings naturally, although it is a time-consuming process. Sometimes they are grown upwards with the help of a trellis or other supporting systems. The hanging-down type (see Figure 3.11) is also a popular vertical landscaping approach. Compared with the wall-climbing type, it can easily form a complete vertical green belt on a multi-storey building through planting at every storey. Lastly, the module type (see Figure 3.12) is a fairly new concept compared with the previous two types. More complicated design and planning considerations are necessary before a vertical system comes into being. In terms of cost, it is probably the most expensive vertical greening method.

Neither natural nor artificial green formations can be isolated in the concrete jungle. Instead, a network of green spaces should be formed. Greenbelt and Greenway are popular land-use planning concepts (Endicott, 1993; Little, 1990; Osborn, 1969; Tan, 2006; Tang et al., 2007) which connect urban and suburban greenery around or within an urban area.

Figure 3.8 Extensive rooftop gardens. (Photos by Yu Chen)

Figure 3.9 Balcony/terrace gardens. (Photos by Yu Chen)

Figure 3.10 Wall-climbing type of vertical landscaping. (Photos by Yu Chen)

Figure 3.11 Hanging-down type of vertical landscaping. (Photos by Yu Chen)

Figure 3.12 Module type. (Photos by Yu Chen)

Table 3.1 Comparison of three vertical landscaping methods.

Type	Plants	Growing media	Construction type
Wall-climbing	Climbing plants	Soil on the ground or in planted box	Minimal supporting structure is needed
Hanging-down	Plants with long hanging-down stems	Soil in planted box on every storey	Planted boxes and supporting structure should be built at according storey
Module	Short plants	Lightweight panel of growing media (such as compressed peat moss)	Supporting structure for hanging or placing modules should be built on façades

Synergistic properties of a network for promoting ecological and human benefits can be achieved in this manner. However, neither concepts really takes into consideration the landscape on buildings. Actually greening buildings should be viewed as a complement to a network of green spaces in a city. It is especially important in tropical cities where hot weather protection in urban spaces and buildings are of equal importance. The involvement of landscape on buildings can expand the current mostly two-dimensional planning concepts to the context of façades and roofs and define a three-dimensional approach to plants in cities.

3.3 Building the green tropical city: the case of Singapore

With an area of 682.7 km² and a population of more than 4 million, Singapore is one of the most densely populated cities in the world. It is difficult to imagine that Singapore Island was fully covered with lush tropical forest in 1819 when Sir Stamford Raffles landed. Dramatic alterations of the land cover have been experienced on the island since then. Over the past 200 years, the island has lost almost 99 per cent of its original forest cover (Corlett, 1992). Singapore is therefore a good example to use to explain the results of rapid deforestation, urbanization and industrialization in tropical areas.

3.3.1 Yesterday

Situated at the southern tip of the Malay Peninsula, Singapore was once fully covered with three major types of tropical forest: around 5 per cent of freshwater swamp forest, 13 per cent of mangrove forest and 82 per cent of tropical rainforest (Chia, Ausafur and Dorothy, 1991; Corlett, 1992; Lu et al., 2005). The coast was overgrown with the mangrove forest and the freshwater swamp forest. The rest of the island was totally covered by the tropical rainforest.

Originally, there were only about 150 habitants living in Singapore. With the establishment of the trading post by Sir Stamford Raffles, the population increased to about 5000 in 1819. Agriculture was introduced into the island when the trade declined in 1834. Introduced economic crops included rubber, nutmeg, pepper, gambier, tapioca and pineapples. The increasing economic benefit drove the rapid deforestation and the total crop area covered about half of the main island. Most valuable timber was depleted although the other half of the island was covered with scattered patches of forest. By 1883, almost 93 per cent of the original forests gave way to agricultural activities (Corlett, 1992; Lu et al., 2005).

In 1882, the government began to pay attention to the issue of rapid deforestation. The forest reserve concept was created and altogether 13 natural reserves with a total area of 4676 ha were formed by 1886. The reserved area increased to 6033 ha in 1907 and reached its highest level at 6579 ha or 11.7 per cent of the land area in 1930. Unfortunately, abandonment of all reserves was proposed afterwards. As a result, the total area of natural reserves dropped to 1940 ha in 1951. Today, there are only three natural reserves left, Bukit Timah (75 ha), Labrador (4 ha) and the Water Catchment Area (2059 ha). Bukit Timah is the only place where the primary rainforest exists. Overall, the continuous development of Singapore has led to the large-scale destruction of the primary vegetation (see Figure 3.13).

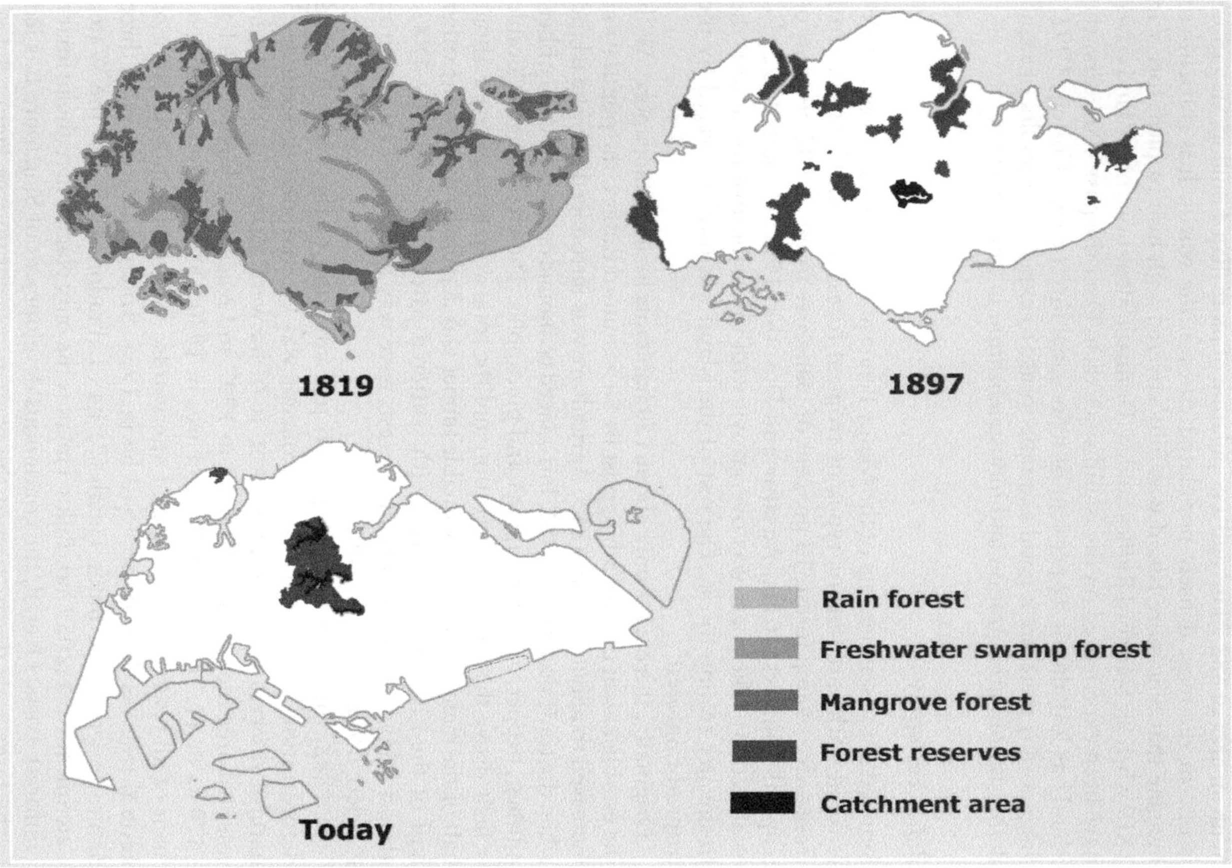

Figure 3.13
Deforestation
in Singapore.
Source: Chia et al.,
1991. (Picture by
Yu Chen)

1897

1819

Today

Rain forest

Freshwater swamp forest

Mangrove forest

Forest reserves

Catchment area

Natural reserves have disappeared to give space for offices, factories and housing developments. At present, only around 6 to 7 per cent of the land surface is for forest and farm-holding areas. Another 42 per cent of the land surface is for inland water, open spaces, public gardens and so on (Chia et al., 1991).

The idea of making Singapore a Garden City originated with the launching of a tree planting campaign in 1967 by the Prime Minister Lee Kuan Yew. Fortunately it has been sustained until today (see Figure 3.14). The purpose of planting was to create a green and beautiful environment. The Park and Recreation Department (PRD) was in charge of creating the image of a Garden City in Singapore. The efforts in landscaping the roads and creating the new urban parks, playgrounds and open spaces made Singapore look really green. Every road has been planted with rows of trees, and every vacant plot of land has been covered by shrubs or grass. Unsightly areas, such as dumping grounds and sub-stations, are screened off with shrubs and climbing plants. The flyovers have been decorated with flowing plants and creepers. Even the lampposts in some parts of Singapore are festooned with climbers. Altogether, more than 4.5 million trees and shrubs have been planted since 1967. There is no doubt that the greening of Singapore is a great success. In addition, according to the concept of the greenway, a proposal to build up a park connector network was approved by the Garden City Action Committee in December 1991 (Tan, 2006). It attempts to link the major parks, nature reserves, natural open spaces and other places of interest in Singapore. The aims are to get more access to parklands and to enhance biodiversity.

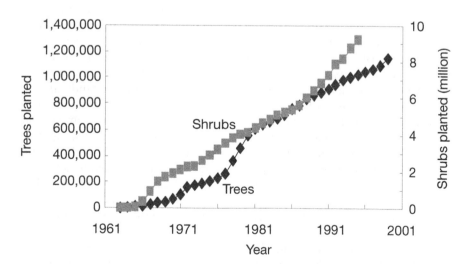

Figure 3.14 Planting in Singapore since 1961.

3.3.2 Today and tomorrow

Today, the lush landscape reflected in the process of planning and developing infrastructure, residential, industrial and commercial buildings has not only beautified the environment but also given great support to Singapore's vibrant economy. The National Parks Board (NParks) is now the agency which is managing the image of the Garden City in Singapore. Currently, there are altogether 1763 hectares of parks, connectors and open spaces (including over 300 parks and playgrounds), 3326 hectares of nature reserves, as well as 4278 hectares of roadside greenery and vacant stateland (source: NParks website http://www.nparks.gov.sg/gardencity_c.asp). The overall estimated quantity of vascular plants, which is a natural part of the Malay Peninsular flora, is around 2277 species in Singapore (Turner, 1993; Turner et al., 1990). However, this is still considered to be relatively poor in the region. With the increasing population and limited land, the Urban Redevelopment Authority (URA) set the target of 0.8 ha of parkland per 1000 persons in 1991 (Lu et al., 2005; Tan, 2006).

However, the management of the current city green spaces is facing some problems. The most significant issue is the fragmentation and species distinction in the nature reserves. For example, the largest Bukit Timah Nature Reserve was divided into two isolated parts by a road, which resulted in discontinuous habitat and constraint of wildlife movement. Meanwhile, the isolated natural reserves are vulnerable to external forces like wind and sunlight (Coops et al., 2004). Besides fragmentation issues, poor health conditions of vegetation in the public green areas is also a concern (Lu et al., 2005).

On the other hand, the short-term benefits of a booming economic environment and the long-term sustainability of an endangered natural environment are always contradicted in the process of rapid urbanization and intensification. Together with the limited land area and natural resources, the urban greening strategies applied in the past 40 years cannot fulfil the requirement of current rapid development, especially when the urban structure extends its limit to a higher level to accommodate more people. The new concept should therefore be developed beyond the landscape on the ground and consider maximizing land-use integration. NParks embarked on the new concept of 'City in a Garden' which is proposed to complement the current 'Garden City' campaign. The new concept still sustains tree planting as well as protection of the existing natural reserves on the island. In addition, the skyrise greenery programme has been launched to achieve a three-dimensional greening instead of the landscape on the ground. The target is to introduce plants into building façades, roofs and walls. NParks believes that creating a three-dimensional garden is not a dream but reality.

Singapore, like many other highly dense cities around the world, has begun to create a three-dimensional garden for our urban environment. Like Chicago, Toronto, Tokyo and Germany, we have incorporated landscaped rooftop gardens and other forms of skyrise greenery in our urban landscapes. In recent years, both private and governmental projects have increasingly reflected these elements.

(http://www.nparks.gov.sg/gardencity/skyrise.shtml)

References

Chia, L. S., Ausafur, R. and Dorothy, T. B. H. (eds) (1991). *The Biophysical Environment of Singapore*. Singapore: Singapore University Press for the Geography Teachers' Association of Singapore.

Coops, N. C., White, J. D. and Scott, N. A. (2004). Estimating fragmentation effects on simulated forest net primary productivity derived from satellite imagery. *International Journal of Remote Sensing*, *25*(4), 819–838.

Corlett, R. (1992). The ecological transformation of Singapore, 1819–1990. *Journal of Biogeography*, *19*, 411–420.

Dirzo, R. and Raven, P. H. (2003). Global state of biodiversity and loss. *Annu. Rev. Environ. Resources*, *28*, 137–167.

Endicott, E. (1993). *Land Conservation through Public/Private Partnerships*. Washington, DC: Island Press.

Fearnside, P. M. and Barbosa, R. I. (2004). Accelerating deforestation in Brazilian Amazonia: towards answering open questions. *Environment Conservation*, *31*, 7–10.

Glantz, M. H. and Krenz, J. (1992). Human components of the climate system. In K. E. Trenberth (ed.), *Climate System Modelling*. Cambridge: Cambridge University Press.

Hansen, M. C. and DeFries, R. S. (2004). Detecting long-term global forest change using continuous fields of tree-cover maps from 8-km Advanced Very High Resolution Radiometer (AVHRR) data for the years 1982–99. *Ecosystems*, *7*, 695–716.

IPCC (2002). *Climate Change 2001: The Scientific Basis*. Cambridge: Cambridge University Press.

Little, C. E. (1990). *Greenways for America*. Baltimore and London: Johns Hopkins University Press.

Lu, X. X., Wong, P. P. and Chou, L. M. (2005). *Singapore's Biophysical Environment*. Singapore: McGraw-Hill.

McGregor, G. R. and Nieuwolt, S. (1998). *Tropical Climatology: An Introduction to the Climates of the Low Latitudes* (Second Edition). New York: Wiley.

Osborn, F. J. (1969). *Green Belt Cities*. London: Evelyn, Adams & Mackay.

Ramankutty, N. and Foley, J. (1999). Estimating historical changes in global land cover: croplands from 1700 to 1992. *Global Biogeochem Cycles*, *13*, 997–1027.

Sodhi, N. S. and Brook, B. W. (2006). *Southeast Asian Biodiversity in Crisis*. Cambridge: Cambridge University Press.

Tan, K. W. (2006). A greenway network for Singapore. *Landscape and Urban Planning*, *76*, 45–66.

Tang, B. S., Wong, S. W. and Lee, A. K. W. (2007). Green belt in a compact city: a zone for conservation or transition? *Landscape and Urban Planning*, *79*, 358–373.

Turner, I. M. (1993). The names used for Singapore plants since 1990. *Gardens' Bulletin*, *45*(1), 1–287.

Turner, I. M., Chua, K. S. and Tan, H. T. W. (1990). A checklist of the native and naturalized vascular plants of the Republic of Singapore. *Journal of the Singapore National Academy of Science, 18 and 19*, 261–273.

Wright, S. J. (2005). Tropical forests in a changing environment. *TRENDS in Ecology and Evolution, 20*, 553–560.

Yamada, I. (1997). *Tropical Rain Forests of Southeast Asia: A Forest Ecologist's View* (P. Hawkes, trans.). Honolulu: University of Hawai'i Press.

Yeang, K. (1998). The skyscraper bioclimatically considered: a design primer. In A. Scott (ed.), *Dimensions of Sustainability: Architecture, Form, Technology, Environment, Culture* (pp. 109–116). London: E. and F. N. Spon.

4 Climate and buildings

4.1 Climate and buildings

In general, climate has a major impact on buildings. This is reflected in the different types of buildings observed in different climatic zones (see Figure 4.1). Temperature, precipitation, humidity, sunlight, wind and other climatic parameters can greatly influence the design and construction of buildings in terms of their layout, orientation, building materials, walls, roofs and so on. As a result, in hot and humid tropical areas, a successful climate-responsive architecture would probably feature a loose layout, big openings, an over-hanging pitched roof and light and thin façade materials. The orientation is decided by the interaction with the sun and the prevailing wind. All these features can respond well to the local hot and humid climate.

Climate also has impacts on buildings in terms of their thermal and visual performances, indoor air quality and building integrity. For instance, a properly oriented building will receive less solar heat gain and result in better thermal performance. A window/opening designed to face the prevailing wind can introduce natural ventilation and create a healthy indoor air quality. Visual performance can be enhanced when daylight is introduced maximally into a building through arranging openings and internal layouts properly. To achieve overall good building performance, it is necessary to consider the climatic impacts on the building performance in a holistic manner instead of focusing on just one or two aspects. On the other hand, climate does not always have a positive impact on building performance. It can be the enemy of building integrity. Erosion, fracture, degradation and weathering of building structures can be caused by rainfall, frost and ice formation, wind and sunshine.

Climate can influence the pattern of energy consumption in buildings as well. It is worth mentioning that this influence was not critical at the very beginning when the traditional human habitats responded well to local climate through employing proper building materials and climate-conscious strategies. However, modern buildings which rely very much upon the utilization of modern mechanical systems consume finite fossil fuels to maintain a comfortable indoor environment. A massive increase in energy

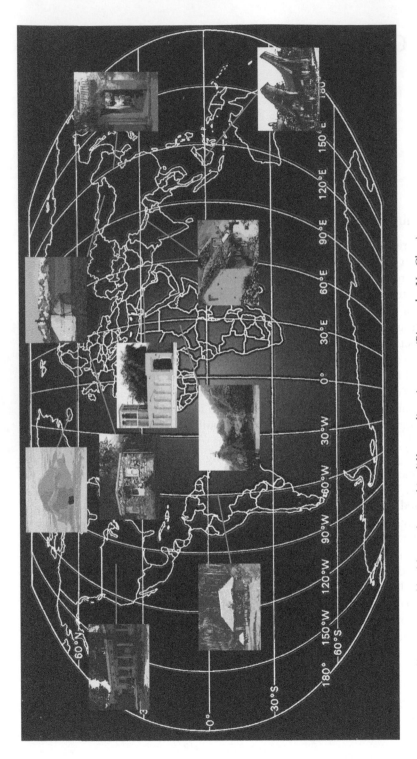

Figure 4.1 Different types of buildings observed in different climatic zones. (Picture by Yu Chen)

consumption began with the era of industrial civilization and it is estimated that roughly 50 per cent of the world fossil fuel consumption is related to the servicing of buildings (Vale and Vale, 1991).

It is not always a one-way influence from climate towards buildings. Buildings, especially in a densely built environment, can in time influence climate. According to Bridgman et al. (1995), buildings in cities influence the climate in five major ways:

1   By replacing grass, soil and trees with asphalt, concrete and glass;
2   By replacing the rounded, soft shapes of trees and bushes with blocky, angular buildings and towers;
3   By releasing artificial heat from buildings, air conditioners, industry and automobiles;
4   By efficiently disposing of precipitation in drains, sewers and gutters, preventing surface infiltration;
5   By emitting contaminants from a wide range of sources, which, with resultant chemical reactions, can create an unpleasant urban atmosphere.

Radiation and heat budgets in a built environment are totally different from those in a natural one. A higher temperature in the centres of cities is therefore observed. This is the well-known urban heat island (UHI) effect. Besides the higher temperature, Bridgman et al. (1995) also believed that lower relative humidity (but higher absolute humidity), higher incidence of fog, lower wind speed (though it is strongly influenced by the orientations of buildings and the street canyons), greater precipitation and cloudiness can be experienced in cities.

4.2 Urban heat island

Rapid urbanization has triggered a number of environmental issues in the built environment. UHI is one of the issues and it is characterized by significantly higher air temperatures in densely built environments as compared to surrounding rural temperatures (see Figure 4.2). The effect was first observed by a meteorologist Luck Howard (1833) in London more than a century ago. Based on the comparison of temperatures of nine years between London and its surroundings, he concluded that the city was warmer at night.

4.2.1 Causes of UHI

The UHI effect has been explored extensively worldwide (Landsberg, 1981; Oke, 1973, 1978, 1982, 1988; Roth et al., 1989; Santamoouris, 2002). According to their studies, the most important factors influencing the UHI effect are summarized as follows (see Figure 4.3):

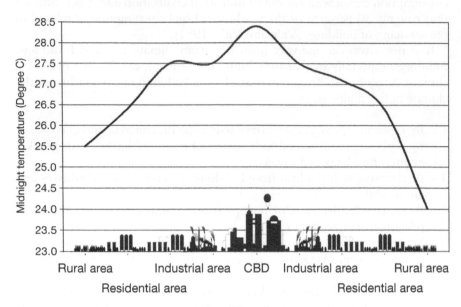

Figure 4.2 Sketch of an urban heat island profile. (Picture by Yu Chen)

1   Canyon geometry
    Urban canyons, especially the deep ones, work as traps which decrease
    the loss of both short-wave and long-wave radiation from the canyons.
    Most incident radiation and long-wave radiation emitted from streets
    and buildings will eventually find their way into indoor space or re-emit
    back to the surroundings after sunset.
2   Building materials
    During the daytime, more sensible heat can be stored in building mater-
    ials, such as concrete, brick and asphalt, due to their big heat capacity.
    The stored heat will be released back to the environment at night.
3   Greenhouse effect
    Long-wave radiation can easily be trapped inside the polluted urban
    atmosphere due to the greenhouse effect.
4   Anthropogenic heat source
    Anthropogenic heat generated from industrial combustion, traffic, air-
    conditioners and so on can aggravate the UHI effect.
5   Evaporative cooling source
    The UHI effect can be mitigated by evaporative cooling means, such as
    vegetation, water body and so on, since more incident energy can be
    transformed into latent heat rather than sensible heat. Unfortunately, the
    lack of such evaporative cooling methods in cities, especially the loss of
    greenery, causes severe UHI effect.

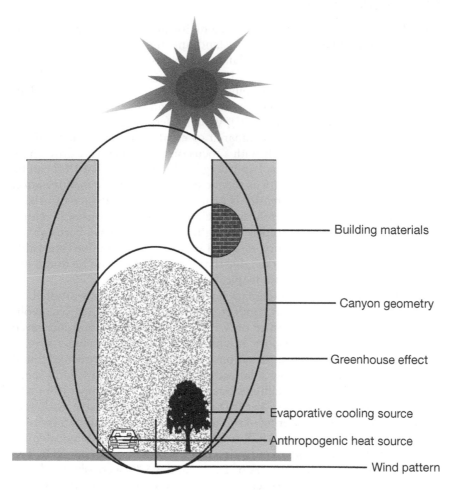

Building materials

Canyon geometry

Greenhouse effect

Evaporative cooling source

Anthropogenic heat source

Wind pattern

Figure 4.3 Diagram showing the most important factors which may influence the severity of the UHI effect. (Picture by Yu Chen)

6   Wind pattern
    Heat trapped inside urban canyons can be advected from source areas by turbulent transfer. However, such heat loss from within streets can be reduced when there is possible obstruction of wind flow by urban settings. In this case, ventilation in urban open space is of great importance.

4.2.2 Intensity of UHI in cities

According to Oke (1978), there are two major atmospheric types of UHI: the urban canopy layer (UCL) heat island and the urban boundary layer (UBL) heat island. The UCL heat island can be observed in the air between the roughness elements (buildings, tree canopies, etc.) while the UBL is located above UCL with a lower boundary subject to the influence of urban surface. In addition, a non-atmospheric heat island type is also recognized, the surface urban heat island (SUHI), which refers to the relatively higher temperature of urban surfaces compared to rural areas. The magnitude of the UHI has been studied mostly with a focus on the temperature differences between rural and urban areas. Landsberg (1981) believed that UHI can be observed in every town and city as the most obvious climatic manifestation of urbanization. The UHI intensity is proportional to the degree of urbanization. The UHI effect is therefore especially obvious in large-scale cities. A number of observational studies throughout the world have been carried out since the UHI effect was first reported by Howard in London (see Table 4.1). It can be observed from the table that the UHI effect can be found not only in temperate cities but also tropical ones.

The hot and humid urban climate has not been extensively studied. In Singapore, the UHI study was conducted as early as the 1960s (Nieuwolt, 1966). The maximum difference of 3.5 °C in temperature was measured between the city and the airport. In another investigation using a similar approach, Chia (1970) took observations on overcast days instead and found out that a combination of low solar radiation receipts and low wind speed together with a low cloud ceiling reduced the city–rural temperature and relative humidity differences. Vertical mixing of air may have occurred, causing a simultaneous lowering of relative humidity and temperatures over all stations monitored. From 1979 to 1981, the Singapore Meteorological Service (MSS) conducted an islandwide survey in order to map out UHI. The maximum temperature differences of 2.5 °C and 4.5 °C were observed on a cool day and a warm day respectively. The difference of around 3 °C was also derived from a field measurement conducted by Tso (1996) in two locations which represented the rural and the urban area respectively in 1992. Nichol (1996) presented UHI in Singapore through remote sensing technology. Over 4 °C difference was observed from the satellite image of Singapore.

4.2.3 How to measure UHI

The UHI effect has been explored in two major ways: ground-based measurements and remote observations. Fixed weather stations or moving vehicles can offer usable data on the ground while satellite and aircraft can provide valuable observations in the air. UHIs have been well defined through the combination of remote sensing and micrometeorology. Most researches presented previously were based on ground measurements. However, remote sensing technologies have been involved in the studies of the UHI extensively

Table 4.1 Intensity of the UHI effect in some European cities

Investigator	Location	Intensity (K)
Chandler (1965)	Kensington and Wisley, UK	3.1
Nieuwolt (1966)	Singapore, Singapore	3–3.5
Oke and Eas (1971)	Montreal, Canada	6.5
Hage (1972)	Edmonton, Canada	6.5
Sani (1973)	Kuala Lumpur, Malaysia	2–2.2 (cloudy day); 4.4–5 (clear day)
Lyall (1977)	London, UK	2.5
Nubler (1979)	Freibourg, Germany	10
Nkederim and Truch (1981)	Calgary, Canada	10.1 (winter); 6 (summer)
Barring et al. (1985)	Malmo, Sweden	7
Swaid and Hoffman (1990)	Essen, Germany	3–4
Adebayo (1990)	Ibadan, Nigeria	3–5
Escourrou (1990/91)	Paris, France	14
Sani (1990/91)	5 cities, Malaysia	3–7
Padmanabhamurty (1990/91)	8 cities, India	0.6–10
Du et al. (1990/91)	Xishuangbanna, P. R. China	>1
Jauregui et al. (1992)	Guadalajara, Mexico	7 (maximum)
Chow et al. (1994)	Shanghai, P. R. China	Up to 8.4
Kuttler et al. (1996)	Stolberg, Germany	6
Yamashita (1996)	Tokyo, Japan	3–8
Tso (1996)	Singapore, Singapore	2.4
Tso (1996)	Kuala Lumpur, Malaysia	4–6.5
Nichol (1996)	Singapore, Singapore	Surface UHI: 4
Klysik and Fortuniak (1999)	Lódz, Poland	2–4; 12 (maximum)
Steinecke (1999)	Reykjavík, Iceland	Very weak
Saaroni et al. (2000)	Tel-Aviv, Israel	3–5
Santamouris et al. (2001)	Athens, Greece	>10
Watkins et al. (2002)	London, UK	2 (mean, daytime); 3.2 (mean, night-time); 8 (maximum)
Rosenzweig et al. (2005)	Newark and Camden, New Jersey, USA	3.0 and 1.5
Wong and Chen (2005)	Singapore, Singapore	4.01
Alexander et al. (2006)	San Juan, Puerto Rico	8 (estimated in 2050)
Tran et al. (2006)	Bangkok, Thailand	surface UHI: 8 (day); 3 (night)
Tran et al. (2006)	Manila, Philippines	surface UHI: 7 (day); 2 (night)
Tran et al. (2006)	Ho Chi Minh City, Vietnam	surface UHI: 5 (day); 2 (night)
Giridharana et al. (2007)	Hong Kong, China	3.4

during the past two decades. According to Voogt and Oke (2003), thermal remote sensors can observe the SUHI since they 'see' the spatial patterns of upwelling thermal radiance. Remote sensing technology can help to explore the UHI effect from the following aspects:

- the spatial structure of urban thermal patterns and their relation to urban surface characteristics;
- the application of thermal remote sensing to the study of urban surface energy balances;
- the application of thermal remote sensing to the study of the relation between atmospheric heat islands and SUHIs.

Except for ground-based measurements and remote observations, a numerical model is also believed to be the method to explore the UHI effect although it relies on the quantitative data derived from the former two methods. Initial studies were based on some two-dimensional modelling (Bornstein, 1975; Estoque and Bhumralkar, 1970). Bornstein (1986) later clarified the usefulness of employing a three-dimensional dynamic model for the study of heat islands and it has been used extensively (Ashie et al., 1999; Kimura and Takahashi, 1991; Saitoh, 1996; Sang et al., 2000; Tanimoto et al., 2004).

4.2.4 Negative impacts of UHI

The UHI effect may have some negative impacts on air quality, human health and cooling energy consumption in cities. First of all, the UHI effect increases the possibility of the formation of smog which is created by photochemical reactions of pollutants in the air. The formation of smog is highly sensitive to temperatures since photochemical reactions are more likely to occur and intensify at higher temperatures. Atmospheric pollution can be aggravated due to the accumulation of smog. In addition, the increased emissions of ozone precursors from vehicles and vegetation are also associated with the high ambient temperature. The UHI effect also involves the hazard of heat-stress related injuries which can threaten the health of urban dwellers.

Higher temperatures in cities will also increase cooling energy consumption and water demand for landscape irrigation. The peak electric demand will be increased as well. As a result, more electrical energy production is needed and this will trigger the release of more greenhouse gas due to the combustion of fossil fuel. The side effects also include the increased pollution level and energy costs. A feedback loop occurs when greenhouse gases eventually contribute to global warming.

The UHI effect is certainly unwelcome in a tropical climate. It induces more energy consumption for cooling throughout the year although it can help to reduce heating energy use in winter in some temperate cities. In addition, since it is highly intensive at night and increases the night-time temperature significantly, the UHI effect will further diminish the small diurnal temperature variation in tropical cities. Urban plan and architectural design may face more challenges in terms of dissipating excessive heat and achieving thermal comfort at night.

4.2.5 Mitigating measures

In order to mitigate the UHI effect in cities, two major strategies are commonly employed, including increasing vegetative cover and using 'cool' materials.

Increasing vegetative cover in cities is the most effective strategy for mitigating the UHI effect. Vegetation, no matter how it is arranged throughout a city in the form of landscapes on the ground or placed around buildings, plays a very important role in regulating the urban climate on different scales (a detailed discussion can be found in Chapters 5 and 6). With its unique 'oasis effect', vegetative cover can modify the energy balance of the whole city by adding more evaporating surfaces. Leaves can seize most of the incoming solar radiation. Except for a very small portion transformed into chemical energy through photosynthesis, most of the incident solar radiation can be transformed into the latent heat which converts water from liquid to gas resulting in a lower leaf temperature, lower surrounding air temperature and higher humidity through the process of evapo-transpiration. Individual buildings can be well shaded by properly placed vegetation and achieve lower surface temperatures throughout the day.

The role of building materials, which is mainly determined by their optical and thermal characteristics, is critical in mitigating the UHI effect as well. Generally, two significant factors, the albedo which is the ratio of the amount of light reflected from a material to the amount of light shining on the material and the emissivity which is the ratio of heat radiated by a substance to the heat radiated by a blackbody at the same temperature, are of equal importance. The former factor governs the absorption of solar radiation and the latter controls the release of long-wave radiation to the surroundings. The so-called 'cool' materials, which are characterized by high reflectivity and high emissivity values, can benefit the urban environment. First, the 'cool' materials can reduce the temperatures of urban hard surfaces through absorbing less incident solar radiation. In addition, less long-wave heat will be released back to the surroundings and lower ambient temperatures can be achieved. Second, because of the lower temperature induced by 'cool' materials, the generation of smog can be reduced. Finally, with 'cool' materials, buildings also have longer lifetimes because they are not easily stressed by the excessive heat.

Besides plants and 'cool' materials, there are some other considerations which may influence the UHI effect in a built environment. One of the main reasons for heat build-up in a city is poor ventilation which can be dominated or modified by urban design. The significant urban design elements include the overall density of the urban area, size and height of the individual buildings, the existence of high-rise buildings and the orientation and width of the streets (Givoni, 1998). Characteristics of canyon geometries, expressed in terms of height-to-width (H/W) and length-to-height (L/H) ratios, are known to produce different air-flow regimes within urban canyons (Oke, 1987).

4.3 Urban ventilation

The study of urban ventilation is highly complex due to the multiple factors affecting the air flow in urban landscape as well as the macro scale of the study. Very often, the studies have to be conducted with simplification of the canyon geometries. The studies can be conducted via the use of wind tunnel facilities (see Figure 4.4) or computational fluid dynamics (CFD) simulations (see Figure 4.5).

Some of the key parameters that have been identified to affect the air-flow in urban landscape (Ng, 2006; Ng et al., 2006) include (see Figure 4.6):

- identification of the prevailing breezeway, air path;
- variation of building height. Varying the height of buildings will significantly improve the penetration of airflow in the urban landscape.

Figure 4.4 Wind tunnel study of airflow in an urban landscape.

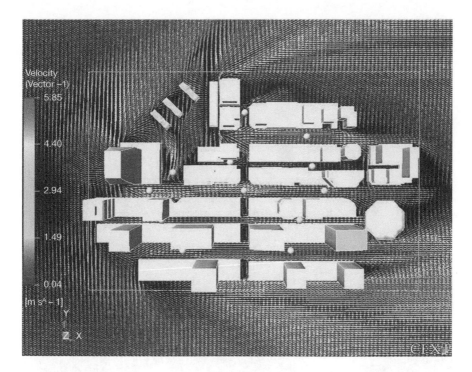

Figure 4.5 CFD simulation of an urban landscape.

Stepping building heights in rows would also create better wind at higher levels if differences in building heights between rows are significant;
• orientation and layout of the buildings/streets with adequate gaps between buildings are essential for good airflow;
• increasing the permeability of building blocks by the provision of void decks at ground level or at mid-span.

4.4 Climate responsive design

According to Hyde (1999), 'climate responsive design is based on the way a building form and structure moderate the climate for human good and well-being'. It may also be termed 'climate-sensitive design' or 'climate-conscious design'. Whichever term is used is insignificant for they all target one objective, which is achieving human comfort in corresponding climatic conditions through the synthesis and the selection of effective climate-related strategies. It is definitely not a new concept but existed in ancient civilizations as illustrated by those vernacular buildings which are compatible with climate through their building forms, fabrics and landscape. However, modern

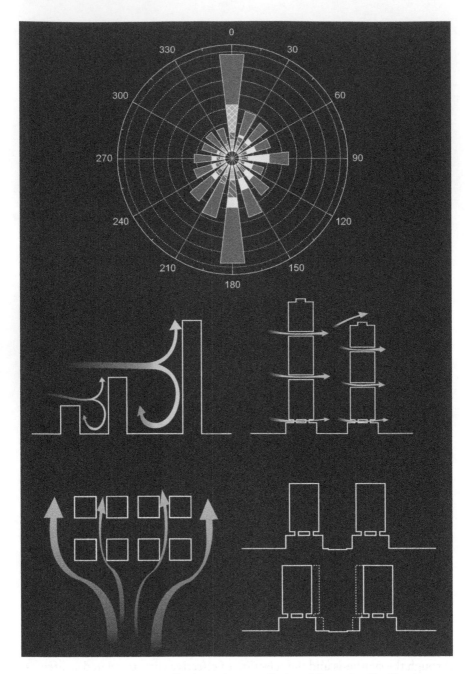

Figure 4.6 Key parameters affecting urban ventilation. (Picture by Yu Chen)

architecture seems to modify the climate in a costly way which largely relies on mechanical plant and equipment in buildings. The recent increasing concern for environmental issues triggered by buildings, however, makes people think about how architecture can respond better to the environment once again. As an integral part of the physical environment, climate has received considerable attention in the context of building design. Respecting the tropical climate is of great importance. Unlike building design in temperate and cold climates with the sense of enclosure, the open and loose layouts of tropical buildings are vulnerable to extreme climatic conditions. More attention should be paid to design in order to modify the climate and create a favourable indoor environment. As early as the eighteenth century, European explorers visited the tropics and realized such differences. Emmanuel (2005) summarized the milestones of the development of the subject in the tropics (see Table 4.2). Fortunately, climate-responsive design has been placed on the agenda after a long period of waning interest in the subject. This is due to the rapid change in the built environment and its related issues globally. It becomes part of the environmental-oriented framework which is seeking sustainable development on the planet. The current expectations for climate-responsive design are not only the saving of finite fossil fuel but the solution for environmental issues such as the UHI effect.

Climate-responsive design in the tropics has been explored extensively (Emmanuel, 2005; Hyde, 1999; Koch-Nielsen, 2002; Koenigsberger et al., 1973; Konya, 1980; Krishan et al., 2001; Olgyay, 1963). It can be simply divided into two parts according to its spatial scale. One is the macro-scale approach which aims to work with climate at the urban level and the other is the micro-scale approach which focuses on the design of individual buildings and their vicinities. The UHI effect can be mitigated by climate-responsive design at both macro- and micro-scales. All the mitigating strategies are the integrated parts under the framework of climate-responsive design (see Table 4.3).

Table 4.2 Summary of tropical climate-sensitive design.

Time	Milestones
1930s	Emerging of the subject of architecture for the tropics
1940s	Study of the deterioration of materials in the tropics
1941	The term 'tropical architecture' first in print
1950s	Study of specific challenges and thermal comfort in the tropics
1960s	Expansion of building types in the tropics
1970s	Textbooks and manuals relating to building types in the tropics
1973	Oil embargo and waning interest in tropical architecture
1980s and 1990s	Barren era of climate-sensitive research and design initiatives
Now	Call for climate-responsive design again due to environmental issues

Table 4.3 Mitigating the UHI effect through climate-responsive design.

Mitigating strategies	Macroscale	Microscale	Objective
Increasing vegetative cover	Introducing parks, gardens, green areas in urban space between buildings	Placing plants around buildings	Increasing evaporative cooling sources and shaded areas
'Cool' materials	Introducing 'cool' material into urban canyons including buildings and pavements	Introducing 'cool' material into roofs and façades	Decreasing solar heat gain during daytime and long-wave heat release at night
Ventilation strategy	Planning density of urban area, the dimension of urban canyons, openness, and the orientation and the width of streets	Orienting buildings, employing openings and sun-shading devices carefully	Avoiding bad solar orientations but encouraging natural ventilation

References

Adebayo, Y. R. (1990). Aspects of the variation in some characteristics of radiation budget within the urban canopy of Ibadan. *Atmospheric Environment, 24B*(1), 9–17.

Alexander, V.-L., Jorge, E. G. and Amos, W. (2006). Urban heat island effect analysis for San Juan, Puerto Rico. *Atmospheric Environment, 40*, 1731–1741.

Ashie, Y., Ca, V. T. and Asaeda, T. (1999). Building canopy model for the analysis of urban climate. *Journal of Wind Engineering and Industrial Aerodynamics, 81*, 237–248.

Barring, L., Mattson, J. O. and Lindovist, S. (1985). Canyon geometry, street temperatures and urban heat island in Malmo, Sweden. *Journal of Climatology, 5*, 433–444.

Bornstein, R. D. (1975). Two-dimensional URBMET urban boundary layer model. *Journal of Applied Meteorology, 14*, 1459–1477.

Bornstein, R. D. (1986). *Urban Climate Models: Nature, Limitation and Application.* WMO Report 652.

Bridgman, H., Warner, R. and Dodson, J. (1995). *Urban Biophysical Environments.* Melbourne and New York: Oxford University Press.

Chandler, T. J. (1965). City growth and urban climates. *Weather, 19*, 170–171.

Chia, L. S. (1970). Temperature and humidity observations on two overcast days in Singapore. *Journal of Singapore National Academy of Science, 1*(3), 85–90.

Chow, S. D., Zheng, J. C. and Wu, L. (1994). Solar radiation and surface temperature in Shanghai City and their relation to urban heat island intensity. *Atmospheric Environment, 28*(12), 2119–2127.

Du, M., Ueno, K. and Yoshino, M. (1990–1991). Heat island of a small city and its influences on the formation of a cold air lake and radiation fog in Xishuangbanna, Tropical China. *Energy and Buildings, 15,* 157–164.

Emmanuel, M. R. (2005). *An Urban Approach to Climate Sensitive Design: Strategies for the Tropics.* London and New York: Spon Press.

Escourrou, G. (1990/91). Climate and pollution in Paris. *Energy and Buildings, 15–16,* 673–676.

Estoque, M. A. and Bhumralkar, C. M. (1970). A method for solving the planetary boundary layer equations. *Boundary-Layer Meteorology, 1,* 169–194.

Giridharana, R., Laua, S. S. Y., Ganesana, S. and Givonib, B. (2007). Urban design factors influencing heat island intensity in high-rise high-density environments of Hong Kong. *Building and Environment, 42,* 3669–3684.

Givoni, B. (1998). *Climate Considerations in Building and Urban Design.* New York: Van Nostrand Reinhold.

Hage, K. D. (1972). Nocturnal temperatures in Edmonton, Alberta. *Journal of Applied Meteorology, 11,* 123–129.

Howard, L. (1833). *Climate of London Deduced from Meteorological Observations* (3rd ed., Vol. 1). London: Harvey and Darton.

Hyde, R. (1999). *Climate Responsive Design: A Study of Buildings in Moderate and Hot Humid Climates.* New York: E. and F. N. Spon.

Jauregui, E., Godinez, L. and Cruz, F. (1992). Aspects of heat-island development in Guadalajara, Mexico. *Atmospheric Environment, 26B*(3), 391–396.

Kimura, F. and Takahashi, S. (1991). The effects of land-use and anthropogenic heating on the surface temperature in the Tokyo metropolitan area. *Atmospheric Environment, 25B,* 155–164.

Klysik, K. and Fortuniak, K. (1999). Temporal and spatial characteristics of the urban heat island of Lódz, Poland. *Atmospheric Environment, 33,* 3885–3895.

Koch-Nielsen, H. (2002). *Stay Cool: A Design Guide for the Built Environment in Hot Climates.* London: James and James.

Koenigsberger, O. H., Ingersoll, T. G., Mayhew, A. and Szololay, S. V. (1973). *Manual of Tropical Housing and Building.* London: Orient Longman.

Konya, A. (1980). *Design Primer for Hot Climates.* London: Architectural Press.

Krishan, A., Baker, N., Yannas, S. and Szokolay, S. V. (eds) (2001). *Climate Responsive Architecture: A Design Handbook for Energy Efficient Buildings.* New Delhi: Tata McGraw-Hill.

Kuttler, W. et al. (1996). Study of the thermal structure of a town in a narrow valley. *Atmospheric Environment, 30*(3), 365–378.

Landsberg, H. E. (1981). *The Urban Climate.* New York: Academic Press.

Lyall, I. T. (1977). The London heat-island in June–July 1976. *Weather, 32,* 296–302.

Ng, E. (2006). Air ventilation assessment system for high density planning and design. Paper presented at the 23rd Conference on Passive and Low Energy Architecture, Geneva, Switzerland.

Ng, E., Wong, N. H. and Han, M. (2006). Permeability, porosity and better ventilated design for high density cities. Paper presented at the 23rd Conference on Passive and Low Energy Architecture, Geneva, Switzerland.

Nichol, J. E. (1996). High-resolution surface temperature related to urban morphology in a tropical city: a satellite-based study. *Journal of Applied Meteorology, 35,* 135–146.

Nieuwolt, S. (1966). The urban microclimate of Singapore. *Journal of Tropical Geography*, 22, 30–37.

Nkederim, L. C. and Truch, P. (1981). Variability of temperatures field in Calgary. *Alberta Atmospheric Environment*, 23–40.

Nubler, W. (1979). Konfiguration and Genese der Warmeinsel der Stadt Freiburg. *Freiburger Geographische Hefte*, 55–63.

Oke, T. R. (1973). City size and the urban heat island. *Atmospheric Environment*, 7, 769–779.

Oke, T. R. (1978). *Boundary Layer Climates*. London: Methuen, New York: Wiley.

Oke, T. R. (1982). The energetic basis of the urban heat island. *Quarterly Journal of the Royal Meteorological Society*, 108, 1–24.

Oke, T. R. (1987). *Boundary Layer Climates* (2nd ed.). London and New York: Methuen.

Oke, T. R. (1988). Street design and urban canopy layer climate. *Energy and Buildings*, 11, 103–113.

Oke, T. R. and Eas, C. (1971). The urban boundary layer in Montreal. *Boundary Layer Meteorology*, 1, 411–437.

Olgyay, V. (1963). *Design with Climate: Bioclimatic Approach to Architectural Regionalism*. Princeton, NJ: Princeton University Press.

Padmanabhamurty, B. (1990/91). Microclimates in tropical urban complexes. *Energy and Buildings*, 15–16, 83–92.

Rosenzweig, C., Solecki, W. D., Parshall, L., Chopping, M., Pope, G. and Goldberg, R. (2005). Characterizing the urban heat island in current and future climates in New Jersey. *Environmental Hazards*, 6, 51–62.

Roth, M., Oke, T. R. and Emery, W. J. (1989). Satellite-derived urban heat islands from three coastal cities and the utilization of such data in urban climatology. *International Journal of Remote Sensing*, 10, 1699–1720.

Saaroni, H., Ben-Dor, E., Bitan, A. and Potchter, O. (2000). Spatial distribution and microscale characteristics of the urban heat island in Tel-Aviv, Israel. *Landscape and Urban Planning*, 48, 1–18.

Saitoh, T. S., Shimada, H. and Hoshi, H. (1996). Modeling and simulation of the Tokyo urban heat island. *Atmospheric Environment*, 30, 3431–3442.

Sang, J., Liu, H. and Zhang, Z. (2000). Observational and numerical studies of wintertime urban boundary layer. *Journal of Wind Engineering and Industrial Aerodynamics*, 87(2–3), 243–258.

Sani, S. (1973). Observations on the effect of a city form and functions on temperature patterns. *Journal of Tropical Geography*, 36, 60–65.

Sani, S. (1990/91). Urban climatology in Malaysia: an overview. *Energy and Buildings*, 15–16, 105–117.

Santamouris, M. (ed.) (2002). *Energy and Climate in the Urban Built Environment*. London: James and James Science Publishers.

Santamouris, M., Papanikolaou, N., Livada, I., Koronakis, I., Georgakis, C. and Assimakopoulos, D. N. (2001). On the impact of urban climate to the energy consumption of buildings. *Solar Energy*, 70(3), 201–216.

Steinecke, K. (1999). Urban climatological studies in the Reykjavik subarctic environment, Iceland. *Atmospheric Environment*, 33, 4157–4162.

Swaid, H. and Hoffman, M. E. (1990). Climatic impacts of urban design features for high and mid latitude cities. *Energy and Buildings*, 14, 325–336.

Tanimoto, J., Hagishima, A. and Chimklai, P. (2004). An approach for coupled simulation of building thermal effects and urban climatology. *Energy and Buildings, 36*, 781–793.

Tran, H., Daisuke, U. b., Shiro, O. b. and Yoshifumi, Y. (2006). Assessment with satellite data of the urban heat island effects in Asian mega cities. *International Journal of Applied Earth Observation and Geoinformation, 8*, 34–48.

Tso, C. P. (1996). A survey of urban heat island studies in two tropical cities. *Atmospheric Environment, 30*(3), 507–519.

Vale, B. and Vale, R. (1991). *Green Architecture: Design for an Energy-conscious Future*. Boston: Little, Brown.

Voogt, J. A. and Oke, T. R. (2003). Thermal remote sensing of urban climates. *Remote Sensing of Environment, 86*(3), 370–384.

Watkins, R., Palmer, J., Kolokotroni, M. and Littlefair, P. (2002). The London heat island – results from summertime monitoring. *BSER and T, 23*(2), 97–106.

Wong, N. H. and Chen, Y. (2005). Study of green areas and urban heat island in a tropical city. *Habitat International, 29*, 547–558.

Yamashita, S. (1996). Detailed structure of heat island phenomena from moving observations from electric tram-cars in metropolitan Tokyo. *Atmospheric Environment, 30*, 429–435.

5 Buildings and plants

5.1 Greenery in a built environment

According to Kiran et al. (2004), 'shrubs, grasses, trees and other forms of natural vegetation are usually the first victims of urbanization'. This indicates the tense relationship between buildings and plants in a built environment. The impact of buildings on urban greenery is definitely not positive at all. First of all, buildings in cities greatly influence the biodiversity of plants since native plants are easily overcome by rapid urbanization. City dwellers may also deliberately or accidentally introduce a large number of exotic plants which compete with the native plants in terms of habitat and resources including water, nutrition and so on. The original habitats for local plants may be cleared or isolated. As a result, many original plants become extinct or endangered in the process of urbanization. The loss of a great number of species in cities means the loss of the ability to self-recover within an eco-system as the number of population interactions within and between species plays an important role in maintaining the health of the system.

In cities, another critical threat is the loss of natural habitats for all plants at a faster pace compared to that in rural areas. The blocky and angular buildings are always replacing the soft shapes of trees, shrubs and grass with asphalt, brick, concrete and glass. Basically, buildings and plants are competitors in terms of space in cities. The truth is that preserving natural habitats and greening cities can never keep up with rapid deforestation and urbanization worldwide.

Besides the loss of biodiversity and natural habitats, plants in cities also face the challenges of rapid urban runoff and severe pollution. The hard surfaces in cities, such as pavements, building façades and roofing, are impervious surfaces which will not allow the absorption of water but channel it rapidly towards cities' storm water discharging systems. The rapid urban runoff means that natural rainfall can not fulfil the water requirement but extra irrigation is needed to maintain the growth of plants in cities. In addition, various chemicals discharged from a built environment concentrate in the atmosphere and the water and can trigger severe pollution. They increase the stress that plants endure in cities. Pollutants in the form of smog and sewage cause sensitive plants to die off easily in cities. Meanwhile, the discharged water which carries the impurities collected from the surfaces of streets, developments and factories can modify the pattern of soil nutrition in suburbs where the destruction of greenery also occurs.

Does this mean that plants should be totally removed from a built environment due to rapid urbanization? The answer is definitely no, since nobody can ignore the benefits brought by plants in a city. Seen from the ecological point of view, flora is the foundation for most ecosystems' food chains and without them, not only would many of the earth's inhabitants perish, but also the earth itself would suffer. In a built environment which is made by humans at the cost of consuming finite natural resources and making massive wastes, the role of plants is equally important in terms of maintaining an ecological balance. Plants together with their related benefits, without any doubt, play an important role in preventing the urban ecosystem from facing its ecological downfall.

5.2 Benefits of greenery in a built environment

Plants in a city can provide quantitative benefits, in the form of financial returns, as well as qualitative environmental, social and aesthetic benefits (see Figure 5.1). Although these benefits are discussed independently later, they are not mutually exclusive and the integrative value of greenery in a built environment should be appreciated whether they are quantitative or qualitative.

5.2.1 Environmental benefits

First of all, plants can offer cooling benefits in a city through two mechanisms, direct shading and evapo-transpiration. The shading effect is quite straightforward and it very much depends on the density of plants. People normally have no quantitative sense of plants' evaporative ability. A good example is that the cooling effect of an isolated mature tree transpiring 450 litres per day from its leaves has been estimated to be equivalent to five average size room air conditioners running 20 hours per day (Pitt, 1979). As a result, not only the shaded hard surfaces but also the ambience can experience relatively low temperatures. The temperature reduction can benefit not only individual buildings but also the urban environment. In other words, the UHI effect can be alleviated through these two simple yet effective mechanisms. This is the focus of this book and the detailed quantitative data related to the thermal benefits of plants in a built environment in the tropics can be found in Chapter 8.

Plants, especially trees, have been widely believed to be effective scavengers of both gaseous and particulate pollutants from the atmosphere in the urban environment (Croxford et al., 1996; Grey and Deneke, 1986; Luley, 1998; MacDonald, 1996; E. G. McPherson, 1998; Miller, 1997; Smith,

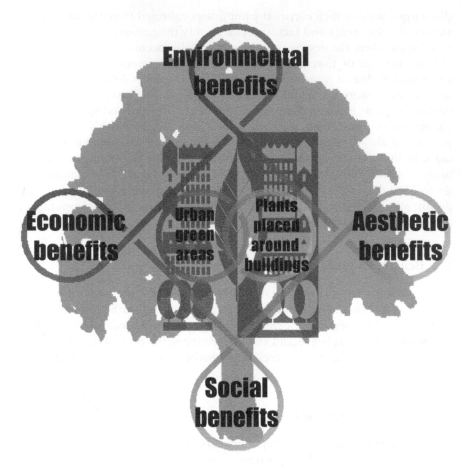

Figure 5.1 Related benefits brought by plants in a built environment. (Picture by
    Yu Chen)

1990). They can improve the air quality by filtering out airborne particles
in their leaves and branches as well as by absorbing gaseous pollutants
through photosynthesis. Inorganic materials (hard surfaces in cities) are not
able to remove pollutants in this manner. For example, it has been estimated
that in Chicago, about 5575 Mg of air pollutants, which include 223 Mg
of carbon monoxide, 706 Mg of sulfur dioxide, 806 Mg of nitrogen dioxide,
1840 Mg of PM_{10} (particles with an aerodynamic diameter of 10 um or
less), and 2000 Mg of ozone, were removed by urban forest in 1991 (E. G.
McPherson et al., 1997; Nowak, 1994). The efficiency of cleansing the air
is decided very much by the species and the amount of urban greenery.
A study in two UK cities (Bealey et al., 2007) showed that reduction of
PM_{10} by 18 to 20 per cent can be achieved when areas have up to a third
of their area available for planting. Plants can also improve air quality

indirectly through mitigating the UHI effect which stirs up dust particles found on the ground. The lower surface temperature caused by plants may also reduce the risk of forming low atmospheric ozone which is the primary component of smog. However, vegetation does not always respond positively to pollution stress. Air pollution has a negative impact on plant metabolism. A reduction of photosynthetic capacity or even the appearance of chlorosis or necrosis can be observed in plants which are planted in a heavily polluted environment (Coleman et al., 1995; Matyssek et al., 1993; Momen et al., 2002). Consideration has also to be given to the species of vegetation since some species can emit volatile organic hydrocarbons (VOCs) that combine with oxides of nitrogen to form smog (Rosenfeld et al., 1997).

Storm water in the urban environment is traditionally routed off impervious surfaces and transported in drainage-pipe systems to an adjacent receiving water body. Flooding may occur in conditions when the drainage system cannot store and subsequently evaporate large amounts of precipitation. Discharging storm water is not an environmentally friendly solution since it is associated with degraded aquatic ecosystems (Miltner et al., 2004; Wang et al., 2001). The ability of the planted surface to retain storm water has often been cited as a practical technique for controlling runoff in a built environment. Environmentally, this translates into benefits such as the reduction of the surface contaminants in the rainwater, reduced occurrence of soil erosion and improved well-being for aquatic plants and animals. Mulching or covering the impervious surfaces with layers of plants and soil is an effective method of conserving water (Adams, 1966). Small plants, such as sedum plants and mosses, are most suitable in this area since they are able to live under extreme conditions with thin growing media and low water supply. A study carried out in Berlin showed that the runoff of a planted area near a railway was only 9 per cent of precipitation while it was up to 70–80 per cent of precipitation for an asphalt surface (Silva et al., 2006). The technique can be applied not only at group level but to impervious roofs. In Berlin, average water retention capacity was observed at 75 per cent of precipitation with 1 m^2 planted modules on a rooftop (Diestel et al., 1993). During the 38th International Federation of Landscape Architects (IFLA) World Congress Conference held in Singapore, a case study (Kohler et al., 2001) on the applicability of green roofs as a storm management system in the humid tropics (Rio de Janeiro) revealed that a green roof allows a significant reduction of the peak load during storm-water events. An annual retention rate of 65 per cent is expected. Several pecularities of the tropics may affect the water retention potential of green roofs. First, frequent storm-water events may result in the potential erosion of newly implemented green roofs and the quicker saturation of the substrate. Secondly, high tropical temperatures may cause a higher evaporation rate and gain in biomass, which presumably increases the water retention rate.

Urban green areas and plants around buildings can be viewed as an acceptable alternative habitat for urban plants and native animals. The presence of wildlife may enrich the ecological quality and health of the environment as well as provide added emotional, intellectual, social and physical benefits to humans (Johnston and Newton, 1996). The size and density of urban greenery decide wildlife benefits. Trees at ground level can provide optimum habitat for insects and birds. Green roofs, with less human intervention, can support sensitive plants and certain animal species, such as butterflies, bees, birds and so on that otherwise might have trouble surviving in urban areas (Brenneisen, 2003; Johnston and Newton, 1996). Briffett (1991) stated that the tropical climate of Singapore has the potential to accommodate and attract resident wildlife and migratory bird life if suitable habitats exist and innovative landscape management policies are implemented. His study proposed incorporating tropical 'plant biodiversity, structural heterogeneity and longitudinal continuity' in the landscape to increase the population and variety of wildlife species.

Plant life can release oxygen to the atmosphere through its unique photosynthesis, which breaks down carbon dioxide and water to create sugars and oxygen. This achieves not only oxygen generation but also carbon dioxide absorption. A study done by Minke and Witter (1982) revealed that one beech tree could produce enough oxygen for ten humans every hour. Carbon emissions in cities can also be cancelled by trees which absorb carbon during the daytime. Meanwhile, plants have the capacity to store carbon in their tissues as they grow. Some studies provide evidence that CO_2 concentrations in urban areas can be 5 to 80 ppm above adjacent rural areas where extensive plants are growing (Berry and Colls, 1990; Clarke and Faoro, 1966; Reid and Steyn, 1997; Takagi et al., 1998). However, both the decomposition of organic matter and the night-time respiration of plants would instead require the consumption of oxygen and the release of carbon dioxide. Plants in cities actually reduce carbon dioxide in an indirect manner rather than a direct one. The ability of vegetation to protect hard surfaces and lower temperature reduces the cooling energy required. It indirectly decreases the emission of carbon dioxide from power plants. Akbari et al. (1990) highlighted the fact that the amount of carbon dioxide avoided via such indirect effects is considerably greater than the amount sequestered directly. It was estimated that the planting of urban trees and the use of light-coloured surfaces can reduce the total carbon production in the United States by about 2 per cent (Akbari et al., 1998).

Plants' roots and the soil can remove some of impurities from the water before it enters a groundwater aquifer. Impurities, such as nitrogen or phosphorus, chemically bond with some types of soil particles. Subsequently, they are removed from the soil and taken up by plants. It is believed that the majority of cadmium, copper and lead as well as notable zinc and nitrogen levels can be taken out of the rainwater by plants (Johnston and Newton, 1996). For example, nitrate concentrations were found to be significantly higher at a suburban site compared to a forested site (Aelion et al., 1997). More researches have also been done on green roofs in the area since water runoff can be easily collected and analysed (Berndtsson et al., 2006; Kohler et al., 2002; Steusloff, 1998). The ability of a green roof to act as a natural filtration mechanism depends on the nature and depth of the layers that make up the system. According to the results obtained from vegetated roofs in Karlsruhe, Germany (Steusloff, 1998), the retention capability for heavy metals depended mainly on the roof's water retention capability. On the other hand, a green roof can also be a possible source of contamination, especially when easily dissolvable fertilizers are used (Berndtsson et al., 2006).

Plants situated between a noise source and a receiver can help to reduce the noise level perceived by the receiver. Since Eyring (1946) first looked into sound attenuation in the jungle, many researchers have investigated the extent to which vegetation could be used to achieve noise control (Bullen and Fricke, 1982; Fricke, 1984; Huddart, 1990; J. Kragh, 1979; U. J. Kragh, 1981; Martínez-Sala et al., 2006; Price et al., 1985). Plants are not totally effective at all frequencies for noise reduction. They generally become transparent when the noise is at frequencies lower than 1 kHz (Bullen and Fricke, 1982). However, it is possible that tree belts are able to attenuate low frequencies more effectively if they are arranged in a lattice configuration (Martínez-Sala et al., 2006). Leaves and branches of the vegetation play an important role in reducing the noise level when it is at the middle frequency range. Tree belts normally take effect at high frequencies (>2 kHz) due to the absorption of sound by the foliage (Fricke, 1984). It has been found that noise reduction can be achieved within a distance of eight times the tree height (Chih and Der, 2005). Green roofs can also provide noise reduction. Hendricks (1994) supported the concept and revealed that the extent to which green roofs can provide acoustical benefits largely depends on the mass of the substrate layer and on existing sound leaks, such as skylights. In a study, green roofs showed their ability to reduce the noise level by up to 50 dB (McMarlin, 1997).

5.2.2 Economic benefits

All economic benefits are closely associated with the environmental benefits brought by plants in a built environment. The ability of vegetative surfaces to retain storm water and lower peak runoff can aid in reducing the extent of storm water drainage infrastructure (Scholz-Barth, 2001). This has been applied by employing smaller storm sewers, which in turn saves construction and maintenance costs of cities' drainage systems. Plants introduced around buildings can improve construction's integrity by lessening the weather effect. According to Peck et al. (1999), the reduction of extreme temperature variation through the use of green roofs can lessen the stress due to expansion and contraction on the membrane, which thereby reduces the cracking and aging of the membrane. The reduced climatic stress on building façades can prolong the service and practical life of buildings. As a result, economic savings due to longer service life, decreased maintenance as well as less replacement can be achieved.

Energy savings are another significant economic contribution brought by greenery in cities. Parker (1983) conducted studies on plants' energy savings in Miami, Florida. He found that the energy savings exceeded 50 per cent on some hot days and the long-term savings were around 25 per cent. McPherson et al. (1988) carried out a similar study which indicated that plant shading on buildings in four US cities can reduce annual cooling costs by 53 per cent to 61 per cent and peak cooling loads by 32 per cent to 49 per cent in a hot climate. Mcpherson and Simpson (1998) also did measurements in Sacramento, California and found that the annual net savings caused by trees for a single building was around $4300. Akbari and Taha (1992) proved that implementing tree planting and cool roofs can achieve large savings in building energy consumption in Canada. The annual savings in heating and cooling costs range from $30 to $180 in urban areas and from $60 to $400 in rural zones. For all building types, over 75 per cent of the total savings were from the direct effects of cool roofs and shade trees (Akbari and Konopacki, 2005). Not only tree shading but also stragetically placed plants around buildings can achieve energy savings. Temperatures in Tokyo could be reduced by 0.11 to 0.84 °C if 50 per cent of its rooftops were planted with vegetation. This would result in energy savings of approximately S$1.6 million per day in electricity bills (Hitoshi, 2000). In Singapore, a hospital has managed to cut its water and electricity bills by S$800,000 in one year after adopting green roofs and other environmental considerations (Nathan, 1999).

Whether residential neighbourhoods or commercial buildings, incorporating green space within their boundaries can easily enhance their values. Therefore, tree planting in cities is a good investment which results in an attractive environment. People are willing to pay more for homes with trees. Studies in America and Britain show that good tree cover increased the property value by 6 to 15 per cent (Peck et al., 1999). Properties with trees have proved to sell or rent faster and better than those without trees (Petit et al., 1995). Strategically placed plants around buildings, in the form of rooftop gardens or vertical landscaping, can also provide similar enhancement of property values. Knepper (2000) revealed that people respond to attractive properties that include socially and environmentally responsible features. Green roofs and vertical landscaping can provide outdoor amenity space and increase the aesthetic appeal of a building, which directly increase the value and marketability of a property. Meanwhile, the high price of urban land may inhibit the development of green spaces. Placing plants around buildings can contribute to the renewal of an urban area.

Plants introduced into buildings, especially rooftop gardens, can help facilitate agricultural production in the urban environment (Kortright and

Hutchinson, 2001). Toronto Food Policy Council (TFPC, 1999) revealed that it is possible to produce a variety of fruit, grain and vegetable crops on rooftops. It is not a dream but reality. In 1997, an urban agricultural firm (Annex Organics) managed to produce saleable tomatoes using an innovative semi-hydroponics system on a roof garden in downtown Toronto (Graneme, 1998). The increasing demand for organic vegetables and the close proximity to the urban area allowed the firm to market its fresh produce successfully. Singapore's Changi Hospital also incorporated hydroponic planters on its rooftop and harvested several crops of leafy greens and fruits, including 150 kg of cherry tomatoes (Nathan, 1999). The viability of cultivation in the urban rooftop may be affected by the presence of urban pollution, the irrigation of plants, the additional structural loading and the potential wind damage. In Japan, NTT Urban Development Corp. and NTT Facilities, Inc. have been jointly testing a sweet-potato hydroponic system on the rooftop of an office building (JapanForSustainability, 2006). Besides the effect of mitigating the UHI effect, sweet potato harvesting was achieved in autumn.

5.2.3 Aesthetic benefits

Landscaping has often been used to improve the aesthetics of the urban environment. The support for the preservation of plants has been attributed to the attraction that many urbanites feel for the natural landscape. Vegetation can provide visual contrast and relief from the highly built-up city environment (Dwyer et al., 1994). Plants also give urban dwellers a significant psychological sense of accessing Mother Nature in concrete jungles where buildings and pavements dominate the landscape. In addition, vegetation provides elements of natural scale and visual beauty as well as a seasonal indicator to buildings and streets. These are some of the amenity benefits brought by plants and they are not easy to be evaluated quantitatively. In addition, incorporating plants around buildings can also offer visual interest and relief to plain walls and roofs and separate them from the obtrusive hard edges of surrounding buildings. Ugly building systems can be hidden by vegetation on the rooftops and façades. The visual appeal from above is an equally important benefit, especially in urban areas with many tall buildings. Residents and workers in high-rise developments often look down on large expanses of ugly asphalt, tiles and slates and flat roofs (Johnston and Newton, 1996).

5.2.4 Social benefits

Plants can fulfil various social functions in a built environment. According to Givoni (1991), plants provide places for playing, sport and recreation, meeting, establishing social contacts, isolation and escape from urban life, aesthetic enjoyment, viewing buildings from a distance and so on. There is no doubt that trees and parks help in creating a sense of community in the neighbourhood. Cities can be made more livable by providing ample amounts of accessible outdoor recreation or amenity space. Green roofs present opportunities for residents to manage communal gardens above ground level. This promotes the feeling of ownership for their roof garden and fosters community interaction. Roof gardens atop office buildings can provide an alternative place for employees to mingle in a more relaxed setting. Given the high premium areas at street level in the cities, roof gardens provide opportunities for more secluded, less polluted and less noisy spaces for informal recreation (Johnston and Newton, 1996).

It has also been proved that visual and physical contact with plants can result in direct health benefits. Ulrich and Parsons (1992) studied the psychological effects of plants on humans and revealed that plants can generate restorative effects leading to decreased stress, improve patient recovery rates and higher resistance to illness. Johnston and Newton (1996) also supported the effect of plants on health and showed that residents having a balcony or terrace garden are less susceptible to illness in a high-density environment. Dr Howard A. Rusk gave some explanations of why horticulture therapy can be successful:

> Some patients are difficult to reach and motivate. Working with plants may provide an impetus and initiate a response . . . One of the great advantages of gardening is that it is not a static activity. There is always something happening – a new sprout, shoot, or leaf is forming, a flower is opening or fading and has to be removed. Then the cycle begins all over again. Most important . . . is a living thing depending on them for care and sustenance. This gives the patient the will to go on and an interest in the future.
>
> (Venolia, 1988: 43)

Dwyer et al. (1994) showed that other 'sensory dimensions' of vegetation can also lead to health improvement. The 'white noise' similar to the sound of wind rustling through leaves has been used in hospital wards to mask disturbing sounds and help heart attack victims to relax. The fragrances of flowers and plants can trigger responses that are more cognitive, and tend to be remembered more vividly than visual stimuli (Porteous, 1985).

Besides their psychological impact, plants have other physical impacts which benefit humans. The air cleansing quality of plants has direct respiratory benefits for people who suffer from asthma and other breathing ailments, and directly lowers smog and other forms of air pollution (Peck et al., 1999). The potential of greenery to lower high temperatures can reduce heat-aggravated illnesses, which directly and indirectly reduce life expectancy of human beings (Weihe, 1986), and death among the city population.

5.3 A new perspective

It is always a dilemma in a city that more buildings should be constructed to meet the requirement of an increased population while more land should be reserved for landscape/greenery to maintain the ecological health of the city at the same time. The German landscape architect Hermann Barges (1986) recommended a new perspective on the greening of cities:

> Towns can be seen as concrete mountains where the streets are ravines and valleys and the houses are like stones or rocks. The roofs of the houses correspond to alpine meadows and pastures. The façades of the houses are slopes, vineyards and terraces. The windows have the appearance of caves and the front gardens are the edges of forests. Courtyards are small valleys; streams are waterways and open spaces in general are deserts and steppes. Finally the chimney and stacks of houses are small volcanoes. If we visualise urban areas in this way we will be able to resettle nature within the town while considering the natural forms we are trying to evoke and thereby how we expect plants to create these illusions for us.

There is no doubt that plants play a significant role in providing a satisfactory environment within cities for their inhabitants. According to Givoni (1991), there are two fundamental categories of greenery in an urban environment: public open space and plants within architectural sites.

Landscape on the ground is still the mainstream of urban green space. Trees along streets, gardens, urban parks, nature reserves and so on are the public green areas. Plants and large public green areas often play a significant role in establishing the image of a city and providing a place where large gatherings and social activities can be carried out. They are considered as environmental luxuries with the increasing population and further urban development.

Plants strategically placed around buildings have become a matter of great interest recently due to their direct benefits to buildings. Balcony gardens, green roofs and vertical landscaping are some good examples. Although greenery placed on hard surfaces cannot totally compensate for the loss of valuable green space on the ground, it is really an ecological approach which can extend the natural environment onto the harsh urban skin – its hard surfaces. Roofs, walls, balconies and other hard surfaces can be transformed into a living landscape, and the ecologically dead areas come alive. All these benefits are so great that they can bring a surprise to even the most enthusiastic advocate of environmentally friendly building (Johnston and Newton, 1996).

In summary, buildings and vegetation should not be the competitors but the collaborators in cities. The importance of vegetation in terms of mitigating the negative impacts caused by rapid urbanization should not be neglected. Greenery should be introduced into the built environment as much as possible in the form of not only large public green areas but also strategically placed items around buildings.

References

Adams, J. B. (1966). Influence of mulches on runoff, erosion and soil moisture depletion. *America Society of Soil Science Proceedings, 30,* 110–114.

Aelion, C. M., Shaw, J. N. and Wahl, M. (1997). Impact of suburbanization on ground water quality and denitrification in coastal aquifer sediments. *Journal of Experimental Marine Biology and Ecology, 213,* 31–51.

Akbari, H., Gartland, L. and Konopacki, S. (1998). Measured energy savings of lightcolored roofs: results from three California demonstration sites. *Proceedings of the 1998 ACEEE Summer Study on Energy Efficiency in Buildings.*

Akbari, H. and Konopacki, S. (2005). Calculating energy-saving potentials of heat-island reduction strategies. *Energy Policy, 33,* 721–756.

Akbari, H. and Taha, H. (1992). The impact of trees and white surfaces on residential heating and cooling energy use in four Canadian cities. *Energy, 17*(2), 141–149.

Akbari, H., Rosenfeld, A. H. and Taha, H. (1990). Summer heat islands, urban trees, and white surfaces. *ASHRAE Proceedings.*

Barges, H. (1986). The city green: Green Vision 2. *Architects' Journal, 6*(183), 40–42.

Bealey, W. J., McDonald, A. G., Nemitz, E., Donovan, R., Dragosits, U., Duffy, T. R. et al. (2007). Estimating the reduction of urban PM10 concentrations by trees within an environmental information system for planners. *Journal of Environmental Management, 85,* 44–58.

Berndtsson, J. C., Emilsson, T. and Bengtsson, L. (2006). The influence of extensive vegetated roofs on runoff water quality. *Science of the Total Environment, 355,* 48–63.

Berry, R. D. and Colls, J. J. (1990). Atmospheric carbon dioxide and sulphur dioxide on an urban/rural transect-1. Continuous measurements at the transect ends. *Atmospheric Environment, 24A,* 2681–2688.

Brenneisen, S. (2003). The benefits of biodiversity from green roofs – key design consequences. Conference on greening rooftops for sustainable communities. Chicago.

Briffett, C. (1991). Environment enhancement techniques in landscape architecture. *Singapore Institute of Architects' Journal, July/August,* 23–27.

Bullen, R. and Fricke, F. (1982). Sound propagation through vegetation. *Journal of Sound and Vibration, 80,* 11–23.

Chih, F. F. and Der, L. L. (2005). Guidance for noise reduction provided by tree belts. *Landscape and Urban Planning, 71,* 29–34.

Clarke, J. F. and Faoro, R. B. (1966). An evaluation of CO_2 measurements as an indicator of air pollution. *Journal of Air Pollution Control Association, 16,* 212–218.

Coleman, M. D., Dickson, R. E., Isebrands, J. and Karnosky, D. F. (1995). Carbon allocation and partitioning in aspen clones varying in sensitivity to tropospheric ozone. *Tree Physiology, 15,* 585–592.

Croxford, B., Penn, A. and Hillier, B. (1996). Spatial distribution of urban pollution: civilizing urban traffic. *Science of the Total Environment, 190,* 3–9.

Diestel, H., Kohler, M. and Schmidt, M. (1993). Funktion begrunter Dacher im stadtischen Raum. *Bundesbaublatt*, 9, 729–734.

Dwyer, J. F., Schroeder, H. W. and Gobster, P. H. (1994). The deep significance of urban trees and forests. In R. H. Platt, R. A. Rowntree and P. C. Muick (eds), *The Ecological City: Preserving and Restoring Urban Biodiversity* (pp. 137–150). Amherst: University of Massachusetts Press.

Eyring, C. F. (1946). Jungle acoustics. *Journal of the Acoustical Society of America*, 18, 257–270.

Fricke, F. (1984). Sound attenuation in forest. *Journal of Sound and Vibration*, 92, 149–158.

Givoni, B. (1991). Impact of planted areas on urban environmental quality: a review. *Atmospheric Environment*, 25B(3), 289–299.

Graneme, S. (1998). *Annex Organics' Rooftop Farming Business*. Retrieved 10 September 2007, from http://cityfarmer.org/rooftopTO.html.

Grey, G. W. and Deneke, F. J. (1986). *Urban Forestry* (2nd ed.). New York: Wiley.

Hendricks, N. A. (1994). Designing green roof systems: a growing interest. *Professional Roofing*, 20–24.

Hitoshi, C. (2000). The sky, the limit. *Look Japan*, 46(535), 6–15.

Huddart, L. (1990). *The Use of Vegetation for Traffic Noise Screening. Transport Research Laboratory Report RR238*. Crowthorne, Berkshire: Transport Research Laboratory.

JapanForSustainability (2006). *Mitigation Effects on the Urban Heat Island Effect by Installing a Sweet-potato Hydroponics System on the Rooftop of an Office Building*. Retrieved 10 September 2007, from http://www.agrometeorology.org/index.php?id=19andbackPID=19andbegin_at=10andtt_news=603.

Johnston, J. and Newton, J. (1996). *Building Green: A Guide for Using Plants on Roofs, Walls and Pavements*. London: The London Ecology Unit.

Kiran, B. C., Mamata, P. and Meena, R. (eds) (2004). *Understanding Environment*. New Delhi and Thousand Oaks, CA: Sage Publications.

Knepper, C. A. (2000). Gardens in the sky. *Journal of Property Management*, 65(2), 36–40.

Kohler, M., Schmidt, M., Grimme, F. W., Laar, M. and Gusmao, F. (2001). Urban water retention by greened roofs in temperate and tropical climates. Conference of the 38th IFLA World Congress. Singapore.

Kohler, M., Schmidt, M., Grimme, F. W., Laar, M., de Assuncao Paiva, V. L. and Tavares, S. (2002). Green roofs in temperate climates and in the hot-humid tropics – far beyond the aesthetics. *Environmental Management Health*, 13(4), 382–391.

Kortright, R. and Hutchinson, T. (2001). *Evaluating the Potential of Green Roof Agriculture*. Retrieved 10 September 2007, from http://www.cityfarmer.org/greenpotential.html.

Kragh, J. (1979). Pilot study on railway noise attenuation by belts of trees. *Journal of Sound and Vibration*, 66(3), 407–415.

Kragh, U. J. (1981). Road traffic noise attenuation by belts of trees. *Journal of Sound and Vibration*, 74, 235–241.

Luley, C. J. (1998). The greening of urban air. *Forum for Applied Research and Public Policy*, 13, 33–35.

MacDonald, L. (1996). Global problems local solutions: measuring the value of the urban forest. *American Forests*, 102, 26–29.

Martínez-Sala, R., Rubio, C., Garcıa-Raffi, L. M., Sanchez-Perez, J. V., Sanchez-Perez, E. A. and Llinares, J. (2006). Control of noise by trees arranged like sonic crystals. *Journal of Sound and Vibration, 291*, 100–106.

Matyssek, R., Keller, T. and Takayoshi, K. (1993). Branch growth and leaf gas exchange of *Populus tremula* exposed to low ozone concentrations throughout two growing seasons. *Environmental Pollution, 79*(1), 1–7.

McMarlin, R. M. (1997). Green roofs: not your garden-variety amenity. *Facilities Design and Management, 16*(10), 32.

McPherson, E. G. (1998). Atmospheric carbon dioxide reduction by Sacramento's urban forest. *Journal of Arboriculture, 24*, 215–223.

Mcpherson, E. G. and Simpson, J. R. (1998). Simulation of tree shade impacts on residential energy use for space conditioning in Sacramento. *Atmospheric Environment, 32*, 69–74.

McPherson, E. G., Herrington, L. P. and Heisler, G. M. (1988). Impacts of vegetation on residential heating and cooling. *Energy and Buildings, 12*, 41–51.

McPherson, E. G., Nowak, D. J., Heisler, G., Grimmond, S., Souch, C., Grant, R. et al. (1997). Quantifying urban forest structure, function, and value: the Chicago Urban Forest Climate Project. *Urban Ecosystems, 1*, 49–61.

Miller, R. W. (1997). *Urban Forestry: Planning and Managing Urban Greenspaces* (2nd ed.). Englewood Cliffs, NJ: Prentice-Hall.

Miltner, R. J., White, D. and Yoder, C. (2004). The biotic integrity of streams in urban and suburbanizing landscapes. *Landscape and Urban Planning, 69*, 87–100.

Minke, G. and Witter, G. (1982). *Haeuser mit Gruenem Pelz, Ein Handbuch zur Hausbegruenung.* Frankfurt: Verlag Dieter Fricke GmbH.

Momen, B., Anderson, P. D., Houpis, J. L. J. and Helms, J. A. (2002). Growth of *ponderosa* pine seedlings as affected by air pollution. *Atmospheric Environment, 36*(11), 1875–1882.

Nathan, D. (1999, 28 May). Hospital's garden feeds patients. *The Straits Times*, Singapore.

Nowak, D. J. (1994). Air pollution removal by Chicago's urban forest. In E. G. McPherson, D. J. Nowak and R. A. Rowntree (eds), *Chicago's Urban Forest Ecosystem: Results of the Chicago Urban Forest Climate Project* (pp. 63–81). Radnor, PA: USDA Forest Service, Northeastern Forest Experimental Station.

Parker, J. H. (1983). Landscaping to reduce the energy used in cooling buildings. *Journal of Forestry, 81*(2), 82–105.

Peck, S. W., Callaghan, C. and Bass, B. (1999). *Greenbacks from Green Roofs: Forging a New Industry in Canada. Status Report on Benefits, Barriers and Opportunities for Green Roof and Vertical Garden Technology Diffusion.* Ottawa: Canada Mortgage and Housing Corporation.

Petit, J., Bassert, D. L. and Kollin, C. (1995). *Building Greener Neighborhoods: Trees as Part of the Plan.* Washington, DC: Home Builder Press.

Pitt, D. (1979). Trees in the City. In I. C. Laurie (ed.), *Nature in Cities: The Natural Environment in the Design and Development of Urban Green Space* (pp. 205–230). Chichester and New York: Wiley.

Porteous, J. D. (1985). Smellscape. *Progress in Human Geography, 9*(3), 356–378.

Price, M. A., Attenborough, K. and Heap, N. (1985). The use of trees for noise control. *Proceedings of Internoise, 85* (pp. 503–506). Munich.

Reid, K. H. and Steyn, D. G. (1997). Diurnal variations of boundary layer carbon dioxide in a coastal city – observations and comparisons with model results. *Atmospheric Environment, 31*, 3101–3114.

Rosenfeld, A. H., Romm, J. J., Akbari, H. and Lloyd, A. C. (1997). *Painting the Town White and Green – MIT Technology Review*. Retrieved 20 October 2001, from http://eetd.lbl.gov/HeatIsland/PUBS/PAINTING/.

Scholz-Barth, K. (2001). *Green Roofs: Stormwater Management from the Top Down*. Retrieved 7 September 2007, from http://www.edcmag.com/CDA/Archives/ d568f635d8697010VgnVCM100000f932a8c0.

Silva, F. O. T., Wehrmann, A., Henze, H.-J. and Model, N. (2006). Ability of plant-based surface technology to improve urban water cycle and mesoclimate. *Urban Forestry and Urban Greening, 4*, 145–158.

Smith, W. H. (1990). *Air Pollution and Forests*. New York: Springer.

Steusloff, S. (1998). Input and output of airborne aggressive substances on green roofs in Karlsruhe. In J. Breuste, H. Feldmann and O. Uhlmann (eds), *Urban Ecology*. Berlin: Springer-Verlag.

Takagi, M., Gyokusen, K. and Saito, A. (1998). Increase in the CO_2 exchange rate of leaves of *Ilex rotunda* with elevated atmospheric CO_2 concentration in an urban canyon. *International Journal of Biometeorology, 42*, 16–21.

TFPC. (1999). *Feeding the City from the Back Forty: A Commercial Food Production Plan for the City of Toronto*. Toronto: Toronto Board of Health.

Ulrich, R. S. and Parsons, R. (1992). Influences of passive experiences with plants on individual well-being and health. In D. Relf (ed.), *The Role of Horticulture in Human Well-being and Social Development: A National Symposium, 19–21 April 1990, Arlington, Virginia*. Portland, OR: Timber Press Inc.

Venolia, C. (1988). *Healing Environments: Your Guide to Indoor Well Being*. Berkeley: Celestial Arts.

Wang, L., Lyons, J. and Kanehl, P. (2001). Impacts of urbanization on stream habitat and fish across multiple spatial scales. *Environmental Management, 28*, 255–266.

Weihe, W. H. (1986). Life expectancy in tropical climates and urbanization. In T. R. Oke (ed.), *Urban Climatology and its Applications with Special Regard to Tropical Areas*. Geneva: World Meteorological Organization.

6 Plants and climate

Climate, especially sunlight, temperature and precipitation, is one of the major ecological forces that govern distribution, abundance, health and functioning of plants. But the extent of the climatic influence varies according to its scale. In return, plants have an influence on the climate. It is necessary to digest the differences between macroclimate, mesoclimate and microclimate before further discussion on climate and plants is made. According to Stoutjesdijk and Barkman (1992):

> Macroclimate, which we may define as the weather situation over a long period (at least 30 yr) occurring independently of local topography, soil type and vegetation ... The mesoclimate, or topoclimate is a local variant of the macroclimate as caused by the topography, or in some cases by the vegetation and by human action ... influences are strongest in the lower 2 m of the atmosphere and the upper 0.5 to 1 m of the soil. The climate in this zone is called microclimate.

6.1 The impact of climate on plants

Without any doubt, the macroclimate governs the distribution patterns of plants all over the world. The soil conditions (e.g. soil development, leaching and podzolization, salt accumulation, erosion by rain and wind, solifluction), which are significant for the growing of plants, are closely related to the localized climate. As a result, the climatic zones on the earth determine what types of plants can survive in the region. Basically, there are two major types of forest in the world: tropical forest as well as temperate and boreal forest. They all strictly follow the climatic boundaries determined by the climate. Figure 6.1 shows the distribution of tropical rainforest in the tropics. To some extent, the macroclimate also shapes the morphology of plants. It has been confirmed by ecologists and plant geographers (Bailey and Sinnott, 1915; Givnish, 1987; Wolfe, 1993) that the increase in the percentage of entire-margined species correlates with increasing temperature and the increase in leaf size correlates with increasing precipitation. This is one of the adaptive features that make plants survive in different habitats in the world, from the extremely cold polar region to the hot and humid tropical area. On the other hand, plants directly exchange water and energy with the environment where some weather elements, such as storms, droughts and floods, severely influence the survival of plants. At present, the question of how global warming will affect plant life has generated much debate. Possible consequences of the global climate change, such as less direct sunlight reaching the earth's surface, more carbon dioxide, higher temperatures and changed rainfall patterns, will heavily influence plants in terms of photosynthesis, frost sensitivity, water-use efficiency, the need for nutrients and so on.

The biological importance of microclimate over plants cannot be ignored either. Each plant has an ideal condition under which it thrives. The condition is normally defined by the availability of sunlight, yearly temperature variations, soil type, soil drainage and water demand which varies according to the microclimate on the spot. Moreover, microclimate governs the heat and water budget, the rate of evaporation and transpiration, the phonological manners, the texture and structure (leaf size, leaf consistency, leaf inclination, etc.) of a plant. Hence, the growth of a plant is very much related to the microclimate under which it is planted.

6.2 The impact of plants on microclimate

According to Koenigsberger et al. (1973), the picture of climate is incomplete without some notes on the character and abundance of plants. Plants do improve the climate, or more accurately, the meso and micro ones by providing a shelter from the sun and wind, decreasing air temperature, increasing humidity and so on. The ability to modify the climate is very much decided by density and species of plants. For example, a meadow can do little to modify the climate as compared to a forest.

In general, plants can adjust climate through their unique shading, wind shielding, evapo-transpiration and photosynthesis processes. Dense foliages can intercept and seize most of the incoming solar radiation. Except for a very small portion transformed into chemical energy through photosynthesis, most of the absorbed solar radiation can be modified to latent heat which converts water from liquid to gas, resulting in lower leaf temperature, lower surrounding air temperature and higher humidity through the process of evapo-transpiration. The whole process can be explained through the energy budget of a plant (Jones, 1991) as follows:

$$\Phi_n - C - \lambda E = M + S$$

where

Φ_n = net heat gain from radiation (short-wave radiation and long-wave radiation). This is often the largest and it drives many other energy fluxes.

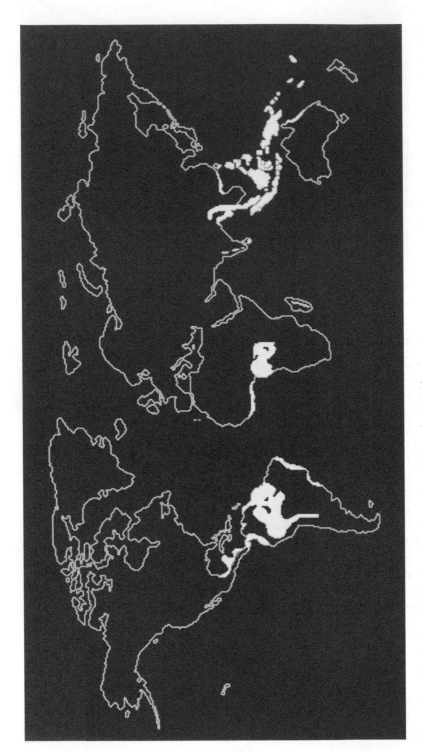

Figure 6.1 Distribution of rainforest in the tropics. (Picture by Yu Chen)

C   = net sensible heat loss, which is the sum of all heat loss to the surroundings by conduction or convection

λE = net latent heat loss, which is that required to convert all water evaporated from the liquid to the vapour state and is given by the product of the evaporation rate and the latent heat of vapourization of water (λ = 2.454 MJ kg $^{-1}$ at 20 °C)

M   = net heat stored in biochemical reactions, which represents the storage of heat energy as chemical bond energy and is dominated by photosynthesis and respiration

S   = net physical storage of thermal energy, which includes energy used in heating the plant material as well as heat used to raise the temperature of the air.

It is necessary to emphasize that energy transferred to latent heat through plants can be very high. For example, an average tree during a sunny day can evaporate 1460 kg of water and consume about 860 MJ of energy (Moffat and Schiler, 1981). On the other hand, any surface covered with plants has a different Bowen ratio, which is the ratio of the sensible heat flux to the latent heat flux, compared to a mineral surface (Barradas et al., 1999). According to Santamouris (2002), the ratio is typically around 5 in a built environment and up to 110 in a desert, while it ranges from 0.5 to 2 in a planted area. A lower Bowen ratio means that lower ambient air temperature can be experienced when similar incident radiation is received by an area. The reason is due to the greater area of leaves which are effective in terms of transforming solar heat gain. Leaf Area Index (LAI) can lie at up to ten times the surface values (Wilmers, 1990/91). Therefore, the oasis effect (opposite to the urban heat island effect) characterized by low ambient air temperature can be observed over an area covered with extensive plants.

    Vegetation can protect against undesired wind as well. A windbreak can be formed when trees, shrubs or other plants are placed perpendicular or nearly so to the principal wind direction. Soil, crop, homestead and road can be protected against the negative effects of wind, such as wind erosion and drifting of soil and snow. Also, plants can be used to redirect the flow of air and channel it to a desired area or location. The ability depends very much on the types of plant. Grassy areas have only a slight effect on sheltering wind while bushes impede wind near the ground. A barrier of dense trees, however, can completely shelter a space from wind to a distance of two or three times their height without creating turbulence.

    In a built environment, the major contributions of plants on urban climate are reflected by their shading and evapo-transpiration cooling effects. Plants can modify the urban climate and benefit buildings through the same process:

1   Reducing solar heat gain through their shading effect;
2   Reducing low-wave heat exchange between shaded surface and the sky;

3   Reducing conductive and convective heat gain through evapo-
     transpiration;
4   Increasing latent cooling through releasing water vapour to the atmos-
     phere.

Thermal protection and comfort can be achieved in this manner for buildings
and urban dwellers. Givoni (1998) highlighted the specific climatic effects of
plants around buildings as follows:

a   Trees with high canopy, and pergolas near walls and windows, provide
     shade and reduce the solar heat gain with relatively small blockage of
     the wind (shading effect).
b   Vines climbing over walls, and high shrubs next to the walls, while
     providing shade, also reduce appreciably the wind speed next to the walls
     (shading and insulation effect).
c   Dense plants near the building can lower the air temperature next to the
     skin of the building, thus reducing the conductive and infiltration heat
     gains. In winter they, of course, reduce the desired solar gain and may
     increase walls' wetness after rains.
d   Ground cover by plants around a building reduces the reflected solar
     radiation and the long-wave radiation emitted towards the walls from
     the surrounding area, thus lowering the solar and long-wave heat gain
     in summer.
e   If the ambient temperature around the condenser of an air conditioning
     unit of a building can be lowered by plants the coefficient of performance
     (COP) of the system can be improved.
f   By reducing the wind speed around a building in winter, plants can
     reduce the infiltration rates and heating energy use of the building
     (insulation effect).
g   Plants on the southern side of a building can reduce its potential to use
     solar energy for heating. Plants on the western and eastern sides can
     provide effective protection from solar gain in summer.

6.3 The impact of plants on solar radiation, wind and humidity

Irradiance reduction due to plants is one of the most direct and effective
impacts by which plants can modify the surrounding climate and protect
buildings easily. The ability of plants to intercept incoming solar radiation
is impressive. Even some leafless deciduous trees in winter can cut off up to
60 per cent or more of irradiance (Heisler, 1986). Higher interception of
radiation, by 70 per cent to 90 per cent, can be achieved by trees with dense
foliage (Chen, 2002; Hoyano, 1988; Papadakis et al., 2001; Simpson, 2002).
The truth that plants prevent most solar radiation from striking surfaces
such as concrete, brick and asphalt implies lower surface temperatures of
shaded areas and lower energy usage for space cooling. Many researches

show that roughly 25 per cent to 80 per cent savings on air conditioning can be achieved with plants strategically placed around buildings (DeWalle et al., 1983; McPherson et al., 1988; Mcpherson and Simpson, 1998; McPherson et al., 1989; Parker, 1983; Raeissi and Taheri, 1999; Simpson, 2002). It is important to emphasize that solar heat extremes can be greatly reduced only when shading is provided at proper locations such as rooftops, and eastern or western exposures which are vulnerable to high intensity of solar radiation. Extra savings can be observed when air-conditioner and central units are well shaded by plants (Petit et al., 1995). From a thermal comfort point of view, the shading effect provided by plants can also reduce the overall radiation absorption by inhabitants through lowering both direct and diffused radiation absorption. Picot (2004) found that an aged tree canopy can generate an energy budget under 50 W/m^2 even with a high ambient air temperature. Meanwhile, unlike mineral hard surfaces, plants reflect less incoming solar radiation back to the environment (Hoyano, 1988).

Carefully placed plants can reduce unfavourable wind velocity. In an open space, an area within up to twelve times the height of a tree belt can be protected from wind forces perpendicular to the plants (Coppin and Richards, 1990). Improvement of microclimate and reduction of wind-loading on buildings can be achieved in this manner. However, plants may play negative roles in reducing wind velocity. Infiltration of outside air, effectiveness of natural ventilation and convective cooling of building surfaces will be influenced (Akbari, 2002). The issue needs extensive attention in the tropics since air movement is an important factor in determining comfort in tropical climates.

The air tends to be more humid when water is slowly released by plants through the process of transpiration to the surrounding environment. This is paired with the reduction of air temperature. The increase of the moisture content of the air can benefit temperate cities where a relatively dry condition is experienced for better comfort (Bernatzky, 1982). However, increased humidity in the tropics may worsen the high-humidity condition in cities. It is generally believed that increased humidity will increase discomfort in hot climates since the loss of metabolic heat via evaporation will be limited and people require a lower temperature for comfort (Ballantyne et al., 1977; deDear et al., 1991; Feriadi and Wong, 2004; Nicol, 2004). On the other hand, there is no conclusive data showing that the increased moisture content in cities is the result of planting. Instead, researchers are inclined to believe that more anthropogenic moisture sources, the increase in urban temperature and the increase in water use in an urban environment, are the possible reasons for observing a 'moisture island effect' (Deosthali, 1999; Emmanuel, 2005; Oke, 1987).

6.4 Temperature reduction caused by plants

It is obvious that plants modify the micro-climate through changing many climatic parameters. But the impact of plants on varying temperatures, either ambient temperature or surface temperature, is well documented. It is simply because temperature is closely associated with the thermal performance of a single building (micro-) or buildings (macro-). In other words, with temperature, the bioclimatic effects of plants can be easily shown and compared in a built environment. This does not mean that the impacts of plants on other climatic parameters are not significant. Instead, their combined effect can be simplified and represented by the variation of temperatures in a built environment.

6.4.1 Studies related to public green areas

Some researches were carried out in exploring the thermal benefits of public green areas. Bernatzky (1982) observed that a small green area in Frankfurt (50–100 m broad) could reduce temperature by up to 3.5 °C and it is equivalent to a geographical rise of the city of 700 m. Meanwhile, relative humidity was increased by 5 to 10 per cent which could ventilate the overheated, dirty and polluted urban area. Jauregui (1990/91) measured Chapultepec Park (around 500 ha) in Mexico City for a period of four years. About 2–3 °C difference on average between the park and its boundaries was observed. The cooling range could reach a distance about the same as its width, which is 2 km. Within the range, the maximum temperature difference obtained at the end of the dry season was up to 4.0 °C although it became smaller during the wet season at only 1 °C (Barradas et al., 1999). Shashua-Bar and Hoffman (2000) explored the cooling effect of small urban greens in Tel-Aviv, Israel. Two significant parameters were found to explain the air temperature variance within the planted areas. These are the partial shaded area under the tree canopy and the air temperature of the adjacent built environment. The average cooling impact could be up to 3 K during the noontime when shading was considered. Shashua-Bar and Hoffman (2004) also predicted the impact of trees within streets and attached courtyards. It has been found that the tree density can govern the cooling effect very much. The cooling effect could be up to 2.5 K and 3.1 K in densely planted streets and courtyards respectively. Other factors, such as cluster deepening, albedo modification and orientation, may also influence the cooling impact of trees. A cooling effect of up to 4.5 K could be achieved when all the factors were correctly arranged.

Many related researches were conducted in Japan. Kawashima (1990/91) explored the effects of vegetation density on the surface temperatures in the urban and suburban areas of Tokyo Metropolis on a clear winter day by use of a satellite image. It has been found that the impact of plants on the surface temperature is marginal in the urban area during the daytime. However, lower surface temperatures could be easily detected in the more vegetated areas in the suburbs during the daytime whereas the inverse situation could be observed at night. In terms of surface temperature variations, fluctuation could be easily observed in the heavily planted areas in the urban environment whereas it was not obvious in the suburbs. Honjo and Takakura (1990/91) developed some numerical models to estimate the cooling impacts of green areas on their surroundings by calculating distributions of temperatures and humidity in Japan. In their calculation, the scale of green areas changed from 100 m to 400 m while their intervals varied from 100 m to 200 m. It has been found that the range of the influenced area near to green areas is a function of the scale of the green areas and the interval between green areas. The authors' conclusion was that effective cooling of surrounding areas by green areas can be achieved by arranging smaller green areas with sufficient intervals. Saito et al. (1990/91) investigated the meteorological data and greenery distribution in Kumamoto City, Japan. In their study, air temperatures were measured by vehicles which can cover relatively large areas. On the other hand, more parameters, such as dry/wet bulb temperatures, globe temperature, wind direction and velocity, were measured in smaller areas. Meanwhile, the land cover and surface temperature were also examined by using remote-sensing data. It has been found that even a fairly small green area (60 m $\times$ 40 m) can impose a cooling effect on surroundings. The maximum temperature difference could be up to 3 °C between the park and its surroundings. The range of cooling area caused by greenery varied with the wind direction. The conclusion was that air temperature distribution in an urban area was closely related to the distribution of greenery. Ca et al. (1998) carried out some field observations to explore the cooling impact of Tama Central Park (0.6 km^2) on surroundings during summertime in the Tama New Town which is at the west of the Tokyo Metropolitan Area. It was found that, at noon, the surface temperatures measured inside the park were 19 °C and 15 °C lower than those measured on the asphalt and the concrete surface respectively. Similarly, air temperature measured at a height of 1.2 m inside the park was more than 2 °C lower than those measured at the same height in the surrounding built environment during noontime. The park acted as a cool source right after sunset since the surface temperatures of the planted area were lower than the air temperatures. In conclusion, with the help of the park, the air temperature in a busy commercial area at 1 km downwind could be reduced by up to 1.5 °C on a hot day. From 1 to 2 pm, a maximum of 4000 kWh of electricity could be saved due to the presence of the park. The cooling energy savings was about US$650.

Due to the combined effect of shade and evapo-transpiration, public green areas in the world can result in an ambient temperature reduction. The temperature difference depends on the sizes of and the distances to the parks. Figure 6.2 shows that temperate reductions of 2 to 3 °C can be achieved by public green areas in some cities. Without any doubt, the reduction of

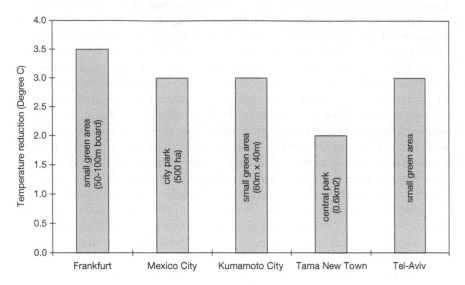

Figure 6.2 The temperature reduction caused by public green areas worldwide.

ambient temperatures can mitigate the UHI effect as well as eventually be translated into cooling energy savings in buildings which are built near to the green areas. From the planning point of view, it can be found that smaller green areas which are strategically arranged or grouped around buildings should be largely promoted in the future. This does not mean that large urban parks are not effective in terms of improving urban climate. But they are considered as a luxury to a heavily built-up environment, especially if rapid urbanization is experienced.

6.4.2 Studies related to rooftop gardens

Regarding the thermal benefits of rooftop gardens, many studies have been done. Harazono (1990/91) developed a vegetation system on the rooftop of a building at the University of Osaka, Japan. In summer, the temperature above the vegetation section was lower than that of the control section by approximately 2 °C. The surface temperature of the rooftop surface without vegetation rose to around 60 °C on clear days, while the peak surface temperature under plants was around 35 °C only. The difference in indoor temperature under planted and exposed roofs was around 2 to 3 °C. The author believed that the whole vegetation system (including vegetation and substrate) should be a good insulation layer for the rooftop. Takakura et al. (2000) observed that the greenery cover had an effective cooling impact on the internal air temperature of some concrete mock-ups. It ranged from around 24 °C to 31 °C throughout the day with plants and 20 °C to 40 °C without plants. Onmura et al. (2001) explored the cooling effect of roof lawn gardens. The surface temperature of the roof slab was reduced by 30 °C by plants during the daytime. Meanwhile, the air temperature beneath the plants was lower than that of the air above by nearly 4 to 5°C. Niachou et al. (2001) measured a green roof in Greece. Surface temperature difference caused by the green roof was of the order of 10 °C. Meanwhile, the indoor thermal condition due to the existence of the green roof was improved by roughly 2 °C. Kumar and Kaushik (2005) developed a mathematical model for evaluating potential cooling impacts of green roofs and solar protection in buildings in Yamuna Nagar, India. They believe average indoor air temperature could be reduced by 5.1 °C compared with that obtained in the room without the green roof.

Green roofs can benefit the urban climate, the microclimate as well as the indoor climate of buildings beneath them (see Table 6.1). Both intensive and extensive rooftop gardens can play a positive role in these areas. However, due to the differences in their constitutions, the performances of the two types of green roofs are not the same. Up to 30 °C reduction of roof surface temperature can be achieved. The ability to reduce the surface temperature is mainly governed by the density or, more accurately, Leaf Area Index (LAI) of plants. Intensive systems are supposed to be better in terms of supporting denser plants, such as shrubs or even trees. Green roofs can also reduce ambient air temperature although the effect is fairly localized according to the studies. Significant differences of around 2 °C can be observed although extremes of up to 4 to 5 °C could be observed underneath the plants. This is reasonable since an intensive rooftop system has denser plants and thick growing media. As part of an integrated system, denser plants can offer a better evaporative cooling impact to the surroundings while a thick growing medium works like extra insulation to building roofs. On the other hand, lower air temperatures indoors and outdoors are achieved by adding extra structural loading and maintenance to the intensive system. Instead of considering thermal benefits alone, potential costs for installation and maintenance of green roofs should be carefully examined. Finally, both surface temperature and air temperature reductions can be translated into energy savings. Actually it is difficult to judge which system is better in terms of energy savings. Both of them are impressive in reducing annual energy consumption or peak space load. In tropical climates, the thermal impact of green roofs is purely positive since they can provide outstanding solar

Table 6.1 The impacts of green roofs on temperatures.

	Temperature reduction (Degree C)
Surface temperature	10 – 30
Ambient air temperature	2
Indoor air temperature	2 – 5.1

protection and insulation to building roofs. However, some researchers also mentioned that green roofs can bring negative savings during wintertime in some temperate climatic regions. This is because the evaporative effect may occur on wet growing media and eventually lead to heat loss from roofs.

6.4.3 *Studies related to vertical landscaping*

Much work has also been done to explore the thermal impact of vertical landscaping. Hoyano (1988) found that a vine sunscreen designed for a south-west veranda could reduce the surface temperature of the veranda floor by 13 to 15 °C and the air temperatures by 1 to 3 °C. A test was also conducted on an ivy-covered west-facing concrete wall. A maximal difference of roughly 18 °C was observed between the walls with and without vegetation. Wilmers (1988) conducted a measurement of some infrared temperatures in Hannover. He also found that the surface temperature of the wall with plants was roughly 10 K below that of the exposed wall in the afternoon. Holm (1989) adapted an hourly computer program to simulate the thermal effect of a deciduous and an evergreen vegetation on an exterior wall. With the climbing plants, the indoor temperature was lowered by 5 °C and was slightly raised at night. Chen (2002) carried out some field measurements and simulations in some residential buildings in Singapore. Maximum surface temperature reduction by 12 °C and air temperature by 1 to 2 °C presented by vertical landscaping were observed in the field measurement. Stec et al. (2005) simulated a double skin façade with plants and found that the temperature of each layer of the double skin façade was much lower for the case with plants than with blinds. The maximum temperature difference between plants and blinds can be up to 20 °C.

Compared with that experienced in green roofs (up to 30 °C), the reduction of surface temperature on the façades of buildings caused by plants are not so obvious (see Table 6.2). This is reasonable since the intensities of solar radiation striking roofs and walls are different. The reduced surface temperatures on building façades, especially on the unfavourable orientations, can still benefit greatly the buildings and the environment. On the one hand, the reduced façade surface temperatures can be translated into cooling energy savings and reduced temperature in buildings. Such energy saving will be significant in tall buildings where the area of vertical wall is greater rather than that of roofs. On the other hand, the reduced surface temperature of

Table 6.2 The impacts of vertical landscaping on temperatures.

	Temperature reduction (Degree C)
Surface temperature	12 – 20
Ambient air temperature	1 – 2
Indoor air temperature	5

the façades will release less heat later to the surroundings and this helps to mitigate the UHI effect.

In this book, the close relationship between plants and climate is discussed in the context of the built environment. How will we ensure the health of our built environment? How will we benefit from plants strategically placed around buildings? How do we ameliorate the UHI effect through greenery? The answers to these questions are based on an understanding of the relationship between plants, climate and buildings. In the tropics, not enough research has been carried out in this area. More quantitative data are necessary before any concrete conclusion can be made.

References

Akbari, H. (2002). Shade trees reduce building energy use and CO_2 emissions from power plants. *Environmental Pollution, 116*, 119–126.

Bailey, I. W. and Sinnott, E. W. (1915). A botanical index of cretaceous and tertiary climates. *Science, 41*, 831–834.

Ballantyne, E. R., Hill, R. K. and Spencer, J. W. (1977). Probit analysis of thermal sensation assessments. *International Journal of Biometeorology, 21*(1), 29–43.

Barradas, V. L., Tejeda-Martinez, A. and Jauregui, E. (1999). Energy balance measurements in a suburban vegetated area in Mexico City. *Atmospheric Environment, 33*, 4109–4113.

Bernatzky, A. (1982). The contribution of trees and green spaces to a town climate. *Energy and Buildings, 5*, 1–10.

Ca, V. T., Asaeda, T. and Abu, E. M. (1998). Reductions in air conditioning energy caused by a nearby park. *Energy and Buildings, 29*, 83–92.

Chen, Y. (2002). An investigation of the effect of shading with vertical landscaping in Singapore. Paper presented at the IFPRA Asia-Pacific Congress 2002, Singapore.

Coppin, N. J. and Richards, I. G. (eds) (1990). *Use of Vegetation in Civil Engineering.* London: Butterworths.

deDear, R. J., Leow, K. G. and Ameen, A. (1991). Thermal comfort in the humid tropics. Part 1: climate chamber experiments on temperature preferences in Singapore. Part 2: climate chamber experiments on thermal acceptability in Singapore. *ASHRAE Transactions, 97*(1), 874–879, 880–886.

Deosthali, V. (1999). Assessment of impact of urbanization on climate: an application of bio-climatic index. *Atmospheric Environment, 33*(24–25), 4125–4133.

DeWalle, D. R., Heilser, G. M. and Jacobs, R. E. (1983). Forest home sites influence heating and cooling energy. *Journal of Forestry, 84*(3), 84–88.

Emmanuel, R. (2005). Thermal comfort implications of urbanization in a warm-humid city: the Colombo Metropolitan Region (CMR), Sri Lanka. *Building and Environment, 40*, 1591–1601.

Feriadi, H. and Wong, N. H. (2004). Thermal comfort for naturally ventilated houses in Indonesia. *Energy and Buildings, 36*(7), 614–626.

Givnish, T. J. (1987). Comparative studies of leaf form: assessing the relative roles of selective pressures and phylogenetic constraints. *New Phytologist, 106*(1), 131–160.

Givoni, B. (1998). *Climate Considerations in Building and Urban Design.* New York: Van Nostrand Reinhold.

Harazono, Y. (1990/91). Effects of rooftop vegetation using artificial substrates on the urban climate and the thermal load of buildings. *Energy and Buildings*, *15–16*, 435–442.

Heisler, G. M. (1986). Energy savings with trees. *Journal of Arboriculture*, *12*(5), 113–125.

Holm, D. (1989). Thermal improvement by means of leaf cover on external walls – a simulation model. *Energy and Buildings*, *14*, 19–30.

Honjo, T. and Takakura, T. (1990/91). Simulation of thermal effects of urban green areas on their surrounding areas. *Energy and Buildings*, *15–16*, 443–446.

Hoyano, A. (1988). Climatological uses of plants for solar control and the effects on the thermal environment of a building. *Energy and Buildings*, *11*, 181–199.

Jauregui, E. (1990/91). Influence of a large urban park on temperature and convective precipitation in a tropical city. *Energy and Buildings*, *15–16*, 457–463.

Jones, H. G. (1991). *Plants and Microclimate* (2nd ed.). Cambridge: Cambridge University Press.

Kawashima, S. (1990/91). Effect of vegetation on surface temperature in urban and suburban areas in winter. *Energy and Buildings*, 15–16, 465–469.

Koenigsberger, O. H., Ingersoll, T. G., Mayhew, A. and Szololay, S. V. (1973). *Manual of Tropical Housing and Building*. London: Orient Longman.

Kumar, R. and Kaushik, S. C. (2005). Performance evaluation of green roof and shading for thermal protection of buildings. *Building and Environment*, *40*, 1505–1511.

McPherson, E. G. and Simpson, J. R. (1998). Simulation of tree shade impacts on residential energy use for space conditioning in Sacramento. *Atmospheric Environment*, *32*, 69–74.

McPherson, E. G., Herrington, L. P. and Heisler, G. M. (1988). Impacts of vegetation on residential heating and cooling. *Energy and Buildings*, *12*, 41–51.

McPherson, E. G., Simpson, J. R. and Livingston, M. (1989). Effects of three landscapes on residential energy and water use in Tucson, Arizona. *Energy and Buildings*, *13*, 127–138.

Moffat, A. and Schiler, M. (1981). *Landscape Design that Saves Energy*. New York: William Morrow and Company.

Niachou, A., Papakonstantinou, K., Santamouris, M., Tsangrassoulis, A. and Mihalakakou, G. (2001). Analysis of the green roof thermal properties and investigation of its energy performance. *Energy and Buildings*, *33*, 719–729.

Nicol, F. (2004). Adaptive thermal comfort standards in the hot-humid tropics. *Energy and Buildings*, *36*, 628–637.

Oke, T. R. (1987). *Boundary Layer Climates* (2nd ed.). London and New York: Methuen.

Onmura, S., Matsumoto, M. and Hokoi, S. (2001). Study on evaporative cooling effect of roof lawn gardens. *Energy and Buildings*, *33*, 653–666.

Papadakis, G., Tsamis, P. and Kyritsis, S. (2001). An experimental investigation of the effect of shading with plants for solar control of buildings. *Energy and Buildings*, *33*, 831–836.

Parker, J. H. (1983). Landscaping to reduce the energy used in cooling buildings. *Journal of Forestry*, *81*(2), 82–105.

Petit, J., Bassert, D. L. and Kollin, C. (1995). *Building Greener Neighborhoods: Trees as Part of the Plan*. Washington, DC: Home Builder Press.

Picot, X. (2004). Thermal comfort in urban spaces: impact of vegetation growth. Case study: Piazza della Scienza, Milan, Italy. *Energy and Buildings, 36*, 329–334.

Raeissi, S. and Taheri, M. (1999). Energy saving by proper tree plantation. *Building and Environment, 34*, 565–570.

Saito, I., Ishihara, O. and Katayama, T. (1990/91). Study of the effect of green areas on the thermal environment in an urban area. *Energy and Buildings, 15–16*, 493–498.

Santamouris, M. (ed.) (2002). *Energy and Climate in the Urban Built Environment*. London: James and James Science Publishers.

Shashua-Bar, L. and Hoffman, M. E. (2000). Vegetation as a climatic component in the design of an urban street. An empirical model for predicting the cooling effect of urban green areas with trees. *Energy and Buildings, 31*, 221–235.

Shashua-Bar, L. and Hoffman, M. E. (2004). Quantitative evaluation of passive cooling of the UCL microclimate in hot regions in summer, case study: urban streets and courtyards with trees. *Building and Environment, 39*, 1087–1099.

Simpson, J. R. (2002). Improved estimates of tree-shading effects on residential energy use. *Energy and Buildings, 34*, 1067–1076.

Stec, W. J., Paassen, A. H. C. v. and Maziarz, A. (2005). Modelling the double skin façade with plants. *Energy and Buildings, 37*, 419–427.

Stoutjesdijk, P. H. and Barkman, J. J. (1992). *Microclimate, Vegetation and Fauna*. Knivsta: OPULUS Press AB.

Takakura, T., Kitade, S. and Goto, E. (2000). Cooling effect of greenery cover over a building. *Energy and Buildings, 31*, 1–6.

Wilmers, F. (1988). Green for melioration of urban climate. *Energy and Buildings, 11*, 289–299.

Wilmers, F. (1990/91). Effects of vegetation on urban climate and buildings. *Energy and Buildings, 15–16*, 507–514.

Wolfe, J. A. (1993). A method of obtaining climatic parameters from leaf assemblages. *U.S. Geological Survey Bulletin 2040*.

7 The plants–climate–buildings model

7.1 A conceptual model

The term 'built environment' is not confined to architecture and buildings but also includes their surrounding environment. From an ecological point of view, buildings are only the man-made subsystems that interact with their environments in a complex manner. To understand and simplify the complexity, some conceptual models have been developed. For example, Yeang (1995) developed a model to interpret the interrelationships between physical and biological constituents in an ecosystem. In his model, structures and buildings are the integrated parts of human communities which interact not only with animal and plant communities but also with climate, geology, soils and hydrologic processes. Yeang believed that, with less interference from human communities, the whole system inclines to be a natural one which can self-maintain and create an inner balance easily. Otherwise the system becomes unstable and faces its downfall when the issues caused by buildings and structures are not solved within the system. On the other hand, to make the man-made system work, a large amount of natural resources should be input from the environment while a similar amount of waste is generated and channelled back to the environment ultimately. The continuous input–output process unavoidably triggers many environmental issues.

In order to find the solution, the focus is therefore shifted from the whole ecosystem to the built environment. Smith et al. (1998) in their book suggested a more complicated model which describes the complex web of the interrelationships in a built environment through adopting models of development, management and planning. Compared to the natural system, plant communities, climate, atmospheric makeup, water quality and other biological components are modified greatly within the physical dimensions of urban areas. An understanding of the modifications, therefore, is the premise for building up a harmony within a built environment and requires successful management.

The preceding two models highlight the role of a built environment in an ecosystem, its surviving mechanism and possible adoptions of biophysical elements, such as plant communities, in the unique artificial system. To acquire a holistic perspective, it is essential to analyse the built environment through involving as many variables as possible. However, it is not easy to manage an urban space by considering every particular aspect presented in the above macro models since 'many of the earth's ecological systems and processes are too complex to be quantified and represented in total' (Yeang, 1995). Instead, 'the crucial task in any theory building is to pick the right variables to be included' (Yeang, 1995). It is realistic to present briefly and simply a range of key components involved and discuss how much interactions/impacts affect the basic system.

Before any key component or model is identified, it is necessary to define the research scope from which the model is deduced. In this case, we should make the environmental issues related to buildings of prime importance. The increasing concern for the environmental impacts of buildings and the quality of their internal environments is believed to be the motivation focusing on these key components and for making such a conceptual model. The proposed model can be placed in the category of environmental control, which has been debated for many years. The Vitruvian tripartite model (see Figure 7.1) is probably the earliest environmental model in architecture. That model simply describes the fundamental relationships between climate, comfort and architecture. Without energy-consumption-based strategies involved to create a favourite indoor environment, the model highlights the nature of environmental control through buildings themselves. The solution is using the fabric of buildings to mediate the external and internal environments. The model stands out as the prototype of the climate-responsive consideration which respects the local climate and emphasizes the importance of building design in the context of façade materials, orientations and so on. However, the lack of consideration of natural forces, like landscape or terrestrial heat, makes the model imperfect.

Olgyay's model can be considered as the successive model (see Figure 7.2) in which the function of 'technology' which consumes energy to create a

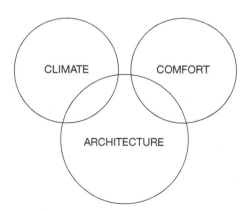

Figure 7.1 Vitruvian tripartite model of environment. (Adapted from Hawkes et al., 2002)

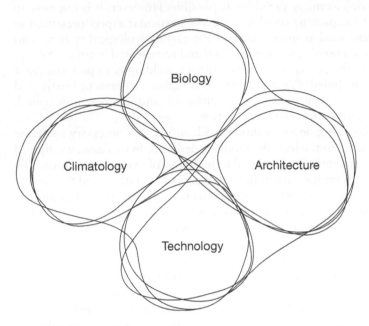

Figure 7.2 Model of environmental process. (Adapted from Olgyay, 1963)

suitable internal environment is taken into consideration. The model also takes into account the functions of 'biology' as well as 'climatology'. In general, the model suggests that an architectural design with respect to 'climatology' and 'biology' is the fundamental consideration while 'technology' is a complementary component which should take effect in the final stage of the mediation. Unfortunately, mechanical heating or cooling now seems to be the most direct way to achieve a 'comfort' indoor environment in any given situation. The emphasis has been placed on technology (energy consumption) rather than biology and climatology in most modern buildings. But Olgyay's model is still the best one which embodies holistic considerations in architectural design.

It is obvious that there are many conceptual models which can guide architectural design from the environmental control perspective. They all share the same goal, that is to reduce environmental impacts caused by the built environment without compromising human well-being. Both the Vitruvian tripartite model and the Olgyay model are good examples in terms of the holistic orientation of environmental design. On the other hand, the general models should be further developed and it is essential to build up specific models which can be the complement or the subset of the general ones. Both the general and the specific models should be considered as integral parts of a complete environmental framework.

Building, climate and plants are highlighted as the key components in this book. Building and climate are the common components which appear in the above general models. These two components are always included in the general models because of the long-term and persistent emphasis on climate control through building designs. Plants, however, are believed to be not only a significant biophysical element but the effective passive method which has been discussed extensively recently. Compiling the three key components into a specific model is meaningful in promoting the passive climate control brought by plants in a built environment. The interactions among the three key components are presented in the model shown in Figure 7.3. The constituent parts of the model are the three key components as well as the interactions (PB, PC, BC and CB). Components with the solid circles indicate relatively stable conditions while the one with the dashed circle implies its potential variation in a built environment. PB is the amount of vegetation introduced into a built environment. This can be enforced when more greenery is introduced into the built environment, such as the reservation of a natural planted area, urban landscaping and strategically introducing plants into buildings. PC is the ability of plants to control the climate. BC and CB are the interactions between climate and buildings. They can be interpreted as the mutual influences, such as the alterations in urban climate caused by cities, the impacts of local climate on the pattern of energy consumption in buildings and so on. The return impacts from climate and buildings towards plants are ignored in the model since the focus is on the built environment rather than on horticulture or botany.

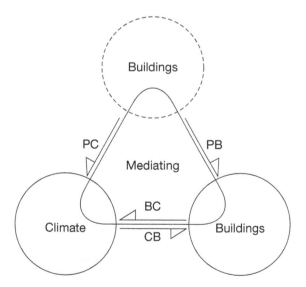

Figure 7.3 Model of environment (plants are considered to be the major component of environmental control). (Picture by Yu Chen)

Based on the model, two fundamental hypotheses can be generated:

$$PB\uparrow \rightarrow PC\uparrow \rightarrow BC\downarrow + CB\downarrow \qquad\qquad \text{Hypothesis 1}$$

$$PB\downarrow \rightarrow PC\downarrow \rightarrow BC\uparrow + CB\uparrow \qquad\qquad \text{Hypothesis 2}$$

where

PB = the amount of vegetation introduced into a built environment
($\uparrow$ increase and $\downarrow$ decrease)
PC = the impact of vegetation over climate control
($\uparrow$ increase and $\downarrow$ decrease)
BC = the negative impact caused by buildings on the urban climate
($\uparrow$ increased and $\downarrow$ decrease)
CB = the negative climatic influence on buildings
($\uparrow$ increase and $\downarrow$ decrease)

The two hypotheses are illustrated in Figure 7.4. They basically imply that the greater the quantity of vegetative intervention upon climate and buildings, the less negative the interactions will be that occur among them and vice versa.

The intervention by plants is very much related to the amount of plants introduced into the built environment as well as their corresponding climate control ability. An optimal and realistic solution is that enough intervention by plants results in acceptable impacts between buildings and climate. Under such circumstances, a balanced urban environment can be created (see Figure 7.5). Otherwise, imbalanced conditions may occur and trigger a

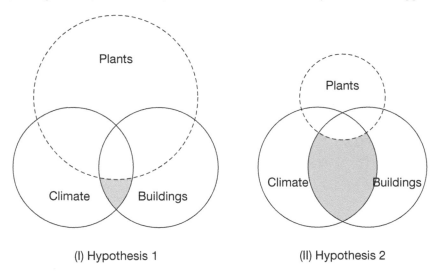

(I) Hypothesis 1   (II) Hypothesis 2

Figure 7.4 Graphical interpretation of the two hypotheses (the shaded area represents the conflicts between climate and buildings). (Pictures by Yu Chen)

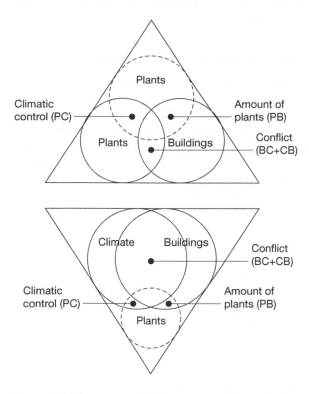

Figure 7.5 The two possibilities generated from the hypotheses – a sustainable possibility (left) and an imbalanced possibility (right). (Picture by Yu Chen)

vicious circle. Bear in mind, sufficient intervention by plants cannot be quantitatively defined by the model at the moment. It relies on individual studies carried out in the different built environments. The challenge is to set the benchmark for both 'sufficient' intervention and 'acceptable' impacts in different climatic conditions.

The plants–climate–buildings model is not a brand-new model but a specific one which can be considered as the complement to the traditional environmental models. The contributions of the model are:

1   Compared with the general environmental models, it inherits environmental concerns but with a greater focus on specific strategy.
2   It can provide both quantitative and qualitative gauges of a building design, especially at the initial stage.
3   The implementation of the model can be carried out for a single building or a built environment as long as the benchmarks can be settled at the different scales.

4 Last but not least, it is expendable when more knowledge regarding the strategy is gained in the future.

7.2 The Garden City movement and its scientific extension

Introducing plants into a built environment is not a new idea. Attempts to do so have been repeated at every stage of human history. Some ancient masterpieces, such as the Hanging Gardens of Babylon, imply the intention of integrating plants in the built environment. But preserving natural landscape in cities was always the only way humans could live with nature in a built environment. Ancient technology could not really set the climatic boundary of city and suburb. Instead, towns were built on top of natural landscape directly and their impact was limited in terms of forming a unique urban climate. It is believed that the image of cities was shaped by the industrial revolution. Since then, there has been an increasing climatic difference between cities and suburbs. Without any doubt, modern solutions based on consuming finite fossil fuel worsens such a contrast by densely packing buildings in cities. Plants have gradually been ruled out of the built environment. People are now realizing their mistake through suffering from related environmental issues. Greening a modern city, to some extent, will not restore original landscape again but can re-introduce plants in forming a new scenario which reflects the use of plants as a passive means to counteract the negative urban impacts.

Ebenezer Howard (1946), the father of the Garden City movement, proposed his idea of making smaller and low-density towns surrounded by belts of undeveloped land. As the pioneers, two representative towns guided by his idea are Letchworth and Welwyn Garden City, in Hertfordshire, England. The idea spread rapidly to Europe and the United States. From the 1920s, the Garden City began booming worldwide. The city of Radburn (New Jersey) was inspired by the success of the two English cities and was developed in the United States by Clarence Stein and Henry Wright. Reconstruction after World War II further stimulated the Garden City movement. Three greenbelt towns (Greenbelt, Maryland; Greenhills, Ohio; and Greendale, Wisconsin) in the United States and the development of over a dozen new communities followed the New Towns Act in 1946 in England, all inherited the characteristics of the Garden City movement. The rise of these Garden Cities has a great influence on the development of modern city planning. Singapore is also a well-known Garden City in the tropical climate. The difference between Singapore and other Garden Cities in Europe and the United States mainly concerns the land use. There is limited undeveloped land in Singapore. But the greening of the city should not be limited to ground level. Instead, the greening of the city should be extended towards façades and roofs of buildings in a more innovative way.

The main purpose of building these Garden Cities is to bring together economic, cultural and ecological benefits. However, the benefits have been taken for granted when the Green City movement is emphasized on the basis of design and planning perspectives alone. There are two omitted yet significant concerns which may need scientific input: how many plants should be introduced and how much the environment will respond. If the two questions remain unanswered, the Green City movement will always be at its infancy stage. Therefore, it is essential that any city to be labelled as a Garden City should be able to demonstrate scientifically the impact of greenery on the urban environment. The conceptual model proposed by the authors is an attempt to answer the two concerns from a particular angle in favour of the further development of the Garden City concept. All the case studies presented in Part II follow the guide of the model in terms of quantitatively defining its four interactions. The benchmarks proposed in the hypothesis are the final goal although it may still need more work in the future.

References

Hawkes, D., McDonald, J. and Steemers, K. (2002). *The Selective Environment*. New York: Spon Press.

Howard, E. (1946). *Garden Cities of Tomorrow*. London: Faber and Faber.

Olgyay, V. (1963). *Design with Climate: Bioclimatic Approach to Architectural Regionalism*. Princeton, NJ: Princeton University Press.

Smith, M., Whitelegg, J. and Williams, N. (1998). *Greening the Built Environment*. London: Earthscan Publications.

Yeang, K. (1995). *Designing with Nature: The Ecological Basis for Architectural Design*. New York: McGraw-Hill.

Part II

8 Case studies in Singapore

In this part, plants, climate and buildings are interpreted by the proposed conceptual model mentioned in Chapter 7 through highlighting the interrelationships among them. In order to test the model with quantitative data, some case studies have been carried out in Singapore which is a green city yet with extensive developments island-wide. There is no distinct boundary between urban and rural areas. However, the two existing major green areas, the primary forest of 75 ha in the middle of the island and the open space allocated for recreation purposes in the northeast of the island, are both located in the northern part of the country, while most built-up regions such as industrial areas, residential areas, the Central Business District (CBD) area and the airport are located in the southern part. Therefore, 'rural' and 'urban' areas in Singapore can be defined in Figure 8.1. We put the model and its two hypotheses in the context of Singapore for the following reasons:

- Singapore is a big modern city which employs high-rise and high-density housing strategy (Buildings);
- Singapore is also a garden city with a large amount of greenery and potential for strategically introducing plants into buildings (Plants);
- Singapore, located in the tropics, has a typical hot and humid climate (Climate);
- Singapore has experienced rapid change of land use during the past few decades which has triggered some environmental issues, such as the UHI effect (Development and conflicts).

The in-depth investigations that follow have been carried out at both micro- and macro levels in order to examine the intervention of greenery in a tropical built environment at the different scales. The detailed available resources and methodologies employed are described as follows:

- satellite images – broad and visible yet instantaneous observation;
- historical weather data – long-term observation;
- mobile survey – observation of a fairly big area within a designated period;

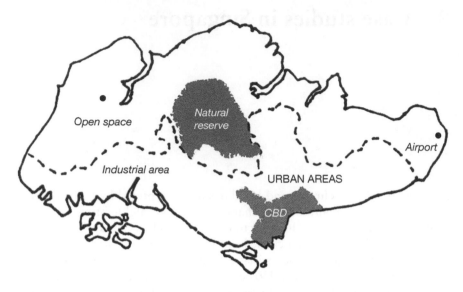

Figure 8.1 The 'urban' and 'rural' partition of Singapore. (Picture by Yu Chen)

- field measurement – close-up observation over specific locations;
- control experiment – observation under designated conditions;
- computational analysis – benefits prediction.

Case study I
UHI measurement

It is undoubted that the UHI effect can be found in almost all cities regardless of their size and it is aggravated mainly due to the loss of green areas in the urban environment. In Singapore, rapid population influx has led to demands for converting natural areas to provide housing and urban infrastructures. In order to have a complete picture, approaching the UHI from different angles was carried out. First, a satellite image with a thermal band was processed to obtain an instantaneous impression of the UHI. The long-term variations accumulated at specific locations were compared subsequently through using historical meteorological data. Finally, island-wide temperature mapping was carried out through mobile surveys.

Satellite image

Introduction

A preliminary study was done to map out the surface temperatures of the entire island of Singapore by the use of satellite images. It aimed to verify the local boundary of urban and rural areas as well as identify the possible hot spots in the island. The Landsat-7 satellite was launched in April 1999, which carries the enhanced thematic mapper plus (ETM+) sensor with the spatial resolutions of 15 m in the panchromatic band, 30 m in the 6 visible bands and 60 m in the thermal band. A Landsat-7 M+ satellite image obtained on 11 October 2002 was selected (see Figure 8.2). The image was radiometrically calibrated and geometrically corrected to obtain the detected radiance data. The detected radiance in the thermal band was converted to equivalent surface temperatures using Planck's blackbody formula. The land cover was classified into several classes, such as vegetation, buildings, roads and so on, using the visible, NIR and SWIR bands. Instead of ambient temperature, the relative surface temperatures can be seen from the satellite image. Within the urban canyon canopy layer in the morning or where wind velocity is low, air temperature is determined by that of adjacent surfaces. Therefore, the surface temperature map is acceptable for studying the UHI effect in the early part of the day (Chen et al., 2001).

Figure 8.2 Landsat-7 ETM+ image of Singapore (acquired on 11 October 2002).

Discussion and observations

The first straightforward observation is the clear surface temperature boundary which coincides with land use in Singapore (see Figure 8.3). The warm region which is represented by red and yellow colour, is mostly located in the southern part of the island where industrial areas, the CBD and airport are located. On the other hand, rural areas are relatively cool with blue and green colour in the northern part of the island. This is due to the concentration of greenery and water bodies as well as less impact from the densely placed urban developments. The clear temperature boundary observed from the image is relevant to the current land-use distribution. The contrast between 'urban' and 'rural' areas hints at the prevalence of the UHI effect in Singapore, although the satellite image only provides the instantaneous observation during the daytime.

The image also reveals some interesting spots with either high or low surface temperatures (see Figure 8.3, see also colour plates). *A* represents some of the industrial areas, such as the south-western Jurong industrial park and the northern Sunget Kadut industrial park. They experienced the highest temperature during the daytime mainly because of the metal roofing employed and the lack of extensive landscape. Similarly, higher temperature was observed in Changi Airport (*B*). This is also reasonable since the exposed

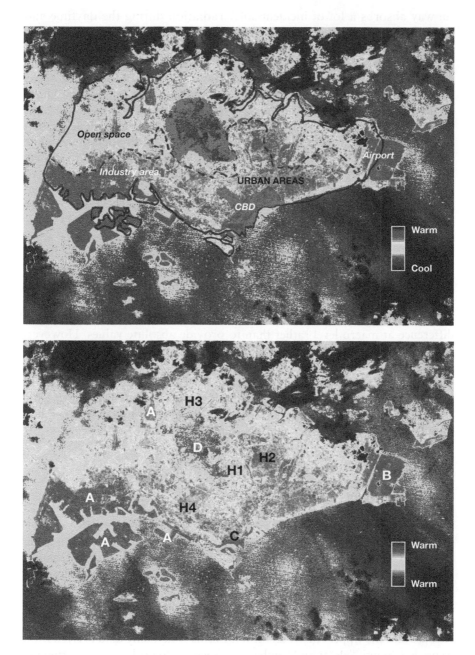

Figure 8.3 Coincidence of surface temperature difference with the boundary of 'rural' and 'urban' areas (top) and hot spots (bottom) in Singapore. (See also colour plates)

runway absorbs a lot of incident solar radiation during the daytime and incurs high surface temperature. *C* represents the CBD area. Compared with the industrial areas and the airport, the temperature in the CBD area was not the worst scenario. Yellow and green can be observed although red dominates in this area. This indicates that the CBD is not the centre of the heat island effect during the daytime. Shadows cast by high-rise buildings in the region can reduce the surface temperatures. For housing estates (*H1* to *H4*), they show slightly different scenarios. *H1* (Toa Poyah new town) and *H3* (Woodlands central) experience relatively low temperature while *H2* (Serangoon new town and Hougang new town) and *H4* (Clementi new town) have slightly high temperatures. *H1* and *H3* are both located near to *D*, the nature reserve and the central catchment area. Therefore, they can probably benefit from the large green area and the water body. *H2* and *H4* are located in the densely built environment which is away from *D* in the urban area. On the other hand, some scattered low-temperature areas can be observed here and there in some Housing and Development Board (HDB) new towns. This is due to the Singapore Government's tree planting campaign which, to some extent, diminishes the urban–rural land cover contrasts.

The reason for using satellite image is to acquire the urban–rural visual difference at macro level rather than to provide absolute values. However, the lack of a traverse observational approach to ascertain the UHI pattern and its intensity as well as time constraints (the satellite image can be taken only during the daytime when the intensity of the UHI is not at its peak) limit the use of satellite imagery for exploring the UHI. In order to trace the gradual influence of urbanization on the local climate, historical weather data obtained from local weather stations were examined.

Historical weather data

Introduction

There are one principal synoptic meteorological station and five supplementary meteorological stations under the Meteorological Services Division (MSD) of the National Environment Agency (NEA). The principal station observes 24 hours each day, while the supplementary stations close for several hours each night. Since 1982, the location of the principal station has been at Singapore Changi Airport. The five supplementary stations are the Paya Lebar Airbase, Tengah Airbase, Seletar Aerodrome, Sembawang Airbase and the Upper Air Observatory. In this study, only data derived from Tengah, Changi, Seletar and Sembawang meteorological stations during the period of 1982 to 2001 were used (see Figure 8.4). Data obtained from Paya Lebar station and the Upper Air Observatory were omitted because non-standard temperature exposure or failure of temperature recording has occurred at the two stations during the past two decades.

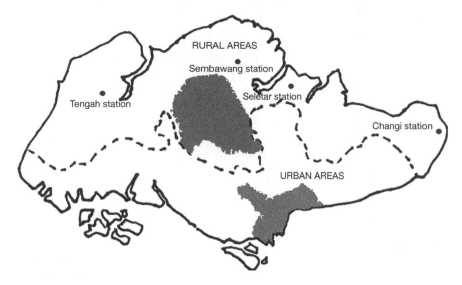

Figure 8.4 Available weather stations in Singapore.

Discussion and observations

The annual mean maximum, mean and mean minimum temperatures obtained at the four stations are given in Figure 8.5. Some important findings are as follows:

1   At Tengah station and Seletar station, the annual maximum, mean and minimum dry-bulb temperatures indicate no significant trend towards warming or cooling during the period 1982–2001, using a 95 per cent confidence level.
2   At Changi station, the maximum, mean and minimum dry-bulb temperatures indicate warming in the annual series. The statistically significant warming trend of each variable in the various series is found to range between 0.03 and 0.05 °C per year for the period 1982–2001.
3   At Sembawang station, the minimum dry-bulb temperatures indicate significant warming annually between 1991 and 2001. The range of this warming trend is between 0.05 to 0.07 °C per year.

In an attempt to further assess the possible effects of urbanization on the local climate, two extreme locations, Tengah and Changi, were explored. Tengah is located within the west region of the main island of Singapore with an area of about 7.4 square kilometres. In 1982, the major land use in Tengah was agriculture with small percentages of temporary residential structures and vacant land. Now its major land use has switched to a military

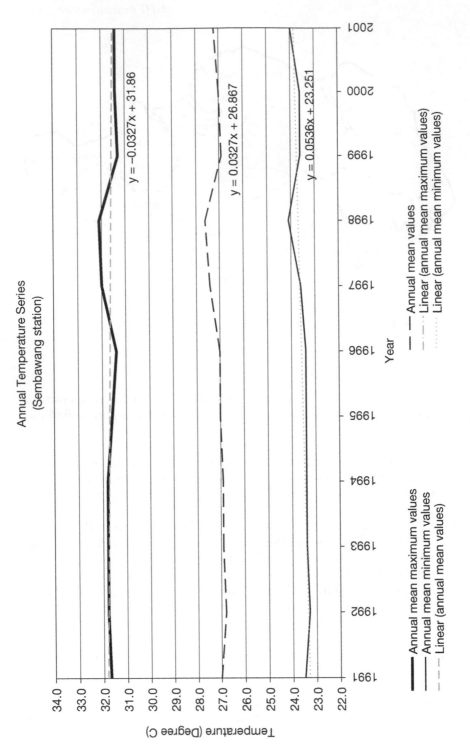

Figure 8.5 Statistical analysis of the past 20 years' weather data.

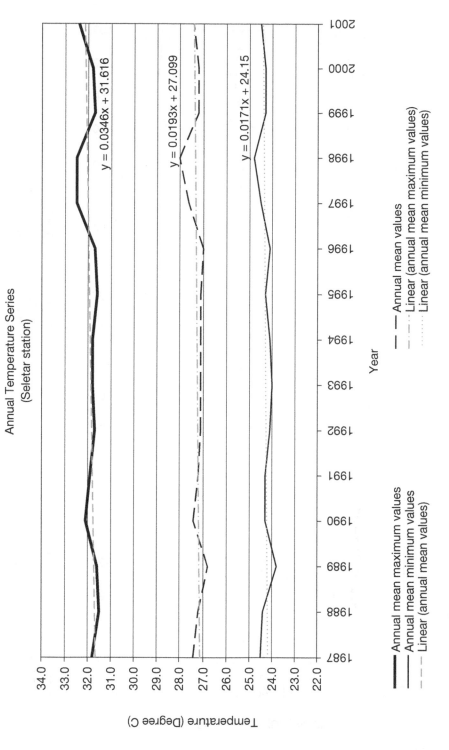

Figure 8.5 Continued

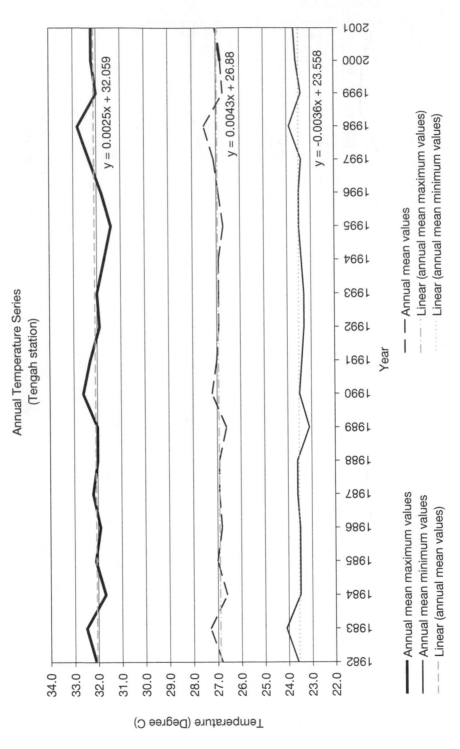

Figure 8.5 Continued

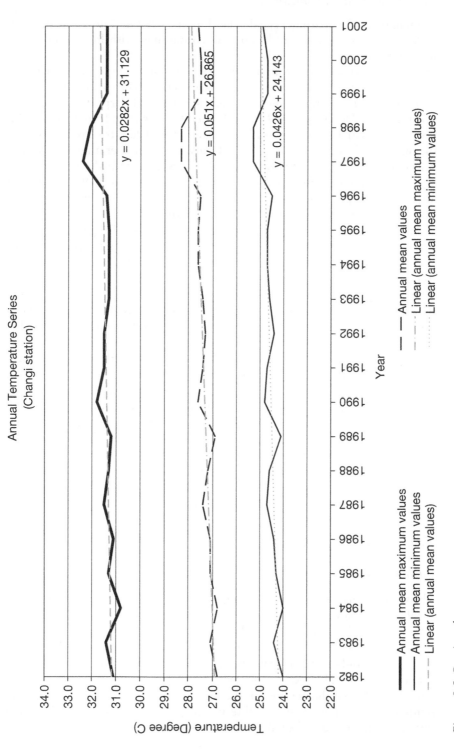

Annual Temperature Series
(Changi station)

$y = 0.0282x + 31.129$

$y = 0.051x + 26.865$

$y = 0.0426x + 24.143$

Year

Temperature (Degree C)

—— Annual mean maximum values
—— Annual mean minimum values
--- Linear (annual mean values)

—— Annual mean values
--- Linear (annual mean maximum values)
······· Linear (annual mean minimum values)

Figure 8.5 Continued

training field. Comparing Tengah's land use in 1982 and now, the physical development of Tengah can be considered minimal with little influence of urbanization.

Changi is located right at the eastern edge of Singapore, covering an approximate area of about 22.7 square kilometres. As early as 1845, it was a fashionable resort for picnics. Changi International Airport was officially opened on 29 December 1981 on reclaimed land and the former British Air Force Base, with one runway and one terminal. In 1982, transportation was the major land use in Changi. A second runway began operating in 1984 and a second terminal in 1990. Now it is also very clear that the predominant land use in Changi is port and airport. In fact, for the period 1982–2001, the airport not only doubled in size when its Terminal 2 opened in 1990, it also saw a significant increase in its annual air traffic volume.

An exploration of the mean temperature trends of Tengah (rural) and Changi (urban) as well as temperature differences between the two locations was made. It can be observed from Figure 8.6 that there is a general upward trend for the Changi site and the urban–rural pair. The differences of mean temperatures for the urban–rural pair, ranging from –0.2 to 1.2 °C, show the rate of increase at 0.05 °C per year. The temperature differences between the two sites fully support the hypothesis that the forming of the UHI is directly related to the urban development. For the period 1982–2001, the planned and implemented physical developments of Tengah and Changi have been very different. Used as an agricultural site in the early 1980s, Tengah has been zoned as a reserve site since the late 1990s and is presently used for military training. As for Changi, the most notable forms of urbanization during the past 20 years would have to be the airport development with its size doubled and its air traffic volume tripled since 1982. The expansion of the airport structure and the increase in air traffic (anthropogenic heat) are the only possible reasons for incurring increased temperatures in Changi Airport. Figure 8.7 provides evidence that the increased temperatures are positively correlated to the air traffic volume at Changi Airport.

Temperature mapping surveys

Introduction

The satellite images show a clear temperature difference between urban and rural areas. However, they can only be used to explain the surface UHI (SUHI) since the satellite 'sees' the spatial patterns of upwelling thermal radiance. On the other hand, the satellite images are consistently taken above Singapore island during the daytime. The UHI effect occurring at night-time cannot be revealed by such remote sensing technology. Although the weather data derived from the network of the local weather stations can remedy the temporal limitation of the satellite images, they are not available in the 'urban' area, especially the CBD district. Some temperature mapping surveys,

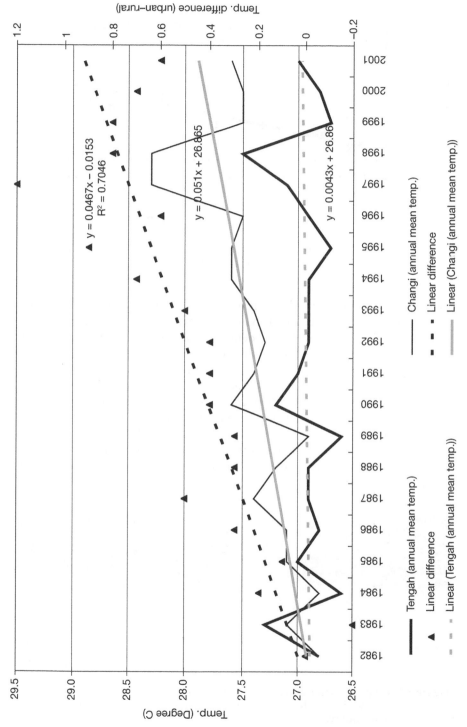

Figure 8.6 Changi–Tengah mean temperature differences.

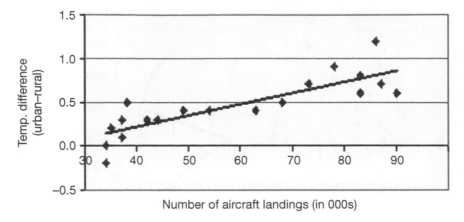

Figure 8.7 Mean temperature difference between Changi and Tengah stations against air traffic volume at Changi Airport.

therefore, have been carried out in order to complement the observations derived from both the satellite images and the historical weather data.

The field survey was done by students holding the observation equipment at a total of 35 designated locations throughout the island on a hot day, 23 August 2002. The measurement covered different land uses such as housing estates, commercial areas and greenery. The measurements were taken simultaneously at all locations in two sessions: the day session between 1 pm and 3 pm and the night session between 9 pm and 11 pm.

The sequent mobile surveys were conducted by vehicles equipped with observation tubes, which can automatically record ambient temperature and humidity at a two-minute measuring interval, at midnight from 2 am to

Figure 8.8 Field measurements carried out at a fixed location and by walking around the area for about 45 minutes at a slow pace to record the temperature.

4 am on 9 July 2002 and 13 September 2002. All vehicles drove along the designated highways at a speed of around 50 Km/h. A single-route survey which crossed the island from the west all the way to the east and a four-routes survey which basically covered the whole island were carried out consecutively (see Figure 8.9).

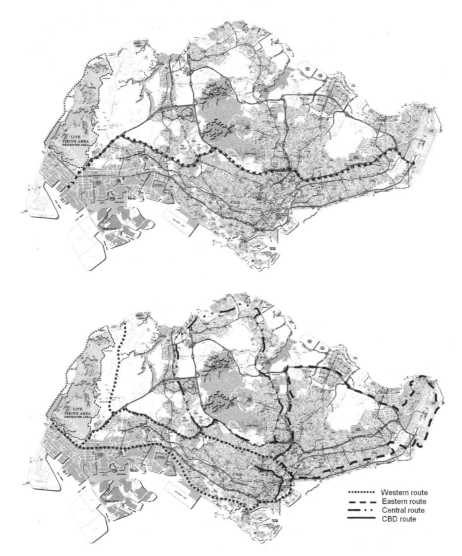

Figure 8.9 A single-route-survey (top) and a four-routes-survey (bottom).

Discussion and observations

a. Field survey

The temperature distribution at various locations during the daytime is shown in Figure 8.10. For every particular location the temperature was calculated by averaging the temperature of all the fixed points. It was observed that the highest temperatures were observed mainly in the industrial areas like Kranji, which is in the southern part, Tuas, which is in the western part of the island and Expo, which is in the eastern part. The industrial areas generally have low-rise buildings and the high temperatures recorded in these areas could be related to the extensive usage of metal roofing in the buildings. The maximum temperature of 33.62 °C was observed at Tuas industrial estate located in the west. Changi Airport also showed a very high temperature of 33.44 °C. This high temperature near the airport can be associated with the large area of concrete runways that absorb a huge part of the solar radiation incident on it and later release it to the atmosphere. The other areas having higher temperatures include mainly the eastern part consisting of Kembangan, Tanah Merah and Paya Lebar. The lowest temperature of 30.78 °C was observed near the Chinese Garden which is a densely vegetated area. The CBD area showed relatively lower temperatures. From the results it can be observed that the daytime temperature seemed to be dominated more by the solar radiation component rather than by the reradiated temperature, which is the main cause of daytime UHI. The shaded area in between tall buildings showed lower temperatures compared to areas with low-rise buildings. The higher temperature in the industrial areas could mainly be due to the metal roofing, which is a commonly used material for industrial buildings in Singapore. All the observations obtained during daytime coincide with those derived from the satellite image.

In urban areas, the night-time temperatures (see Figure 8.11) varied between 27.36 °C and 30.31 °C and it was found that the CBD area was about 2 °C hotter than the locations with greenery. This also indicates the centre of the night-time UHI effect which has shifted from the industrial areas during the daytime to the CBD area. The average temperature of 29.27 °C was recorded around the Orchard area, whereas the area near the Macritchie Reservoir, which is a large green area located in the central part of the island, showed an average temperature of 27.36 °C. The maximum temperature of 30.31 °C was still noted near Changi Airport. Other locations which showed higher temperatures (above 29 °C) include Raffles Place, Outram Park, Clementi and Toa Payoh, all of which comprise densely populated buildings. Places like Kembangan and Paya Lebar also had temperatures above 29 °C. It was found that Woodlands, Sembawang and Kranji situated in the northern part of the island showed relatively lower temperatures.

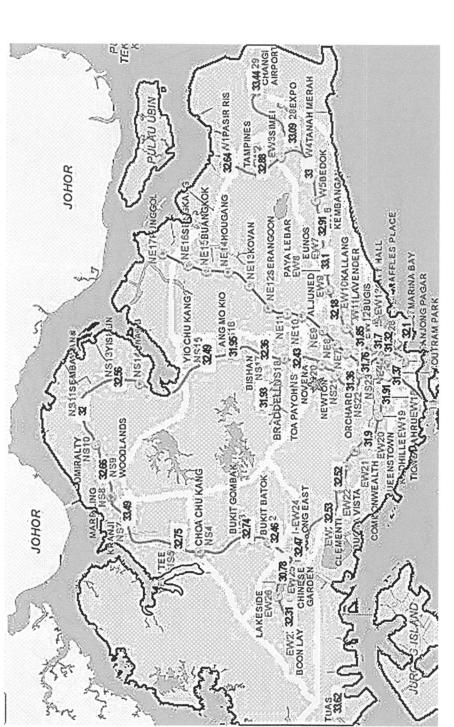

Figure 8.10 The average temperature measured at every MRT station during the daytime (1300–1500 hr).

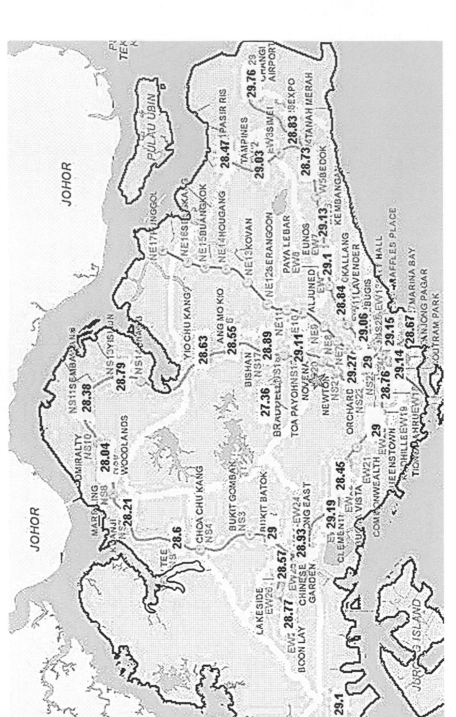

Figure 8.11 The average temperature measured at every MRT station during the night-time (2100–2300 hr).

b. Mobile survey

The one-route survey was to preliminarily investigate the correlation between temperatures/relative humidity and different land uses. The route running from west to east actually passed through quite a number of different land uses, like industrial areas, residential areas, the forest and the airport. The results are illustrated in Figure 8.12. The solid line shows the fluctuation of ambient air temperatures while the dotted line represents the variation of relative humidity. The highest air temperature, 28.6 °C, was observed in the Bedok industrial area at around 2:43 am. The lowest temperature, 27.0 °C, was recorded when the car passed through the southern edge of the natural reserve and the catchment area at around 2:28 am. The Changi Airport and Tuas industrial area also reached relatively high temperatures, around 28.3 °C to 28.5 °C while some residential areas like Bukit Batok and Toa Payoh had lower temperatures around 27.7 °C. The fluctuation of relative

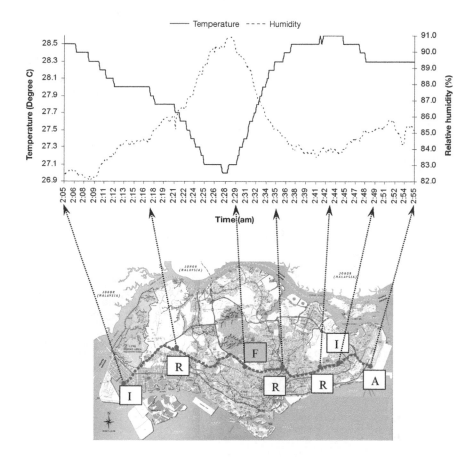

Figure 8.12 The results of the one-route survey (I = Industry area; R = Residential area; F = Forest; A = Airport).

humidity took the opposite direction to the temperature profile. The highest relative humidity of 90.9 per cent was observed near the forest while the lowest relative humidity of 82.1 per cent was obtained near an industrial area. The maximum difference of ambient air temperature obtained in the survey was 1.6 °C. To some extent, this indicates the clear correlation between temperature variation and land uses at night.

The temperature distribution obtained from the four-route survey, which covers the whole island, is shown in Figure 8.13. It can be found that the lower temperatures were mostly detected in the northern part while higher temperatures were observed in the southern part, especially in the CBD area. The results of the survey accorded with the partition of 'urban' and 'rural' areas in Singapore. The lowest temperate, 24.3 °C, was observed in the north-western open space/recreation area where vegetation was well planted. It is estimated that even lower temperatures could be obtained inside the primary forest. The highest temperature, 28.31 °C, was detected in the high-rise and high-density CBD area where less vegetation was planted. Maximally around 4 °C temperature difference was detected in this second mobile survey.

Based on the data derived from the mobile survey, a sketch of the UHI profile in Singapore was plotted (see Figure 8.14). It reveals that the temperatures measured within the different land uses are quite relevant to the existence of greenery. In the CBD area, the absence of plants and high density of buildings caused the highest temperature. Higher temperatures were also detected in some industrial areas. The possible reason may be due to the employment of metal roofs for industrial buildings and lack of tall trees

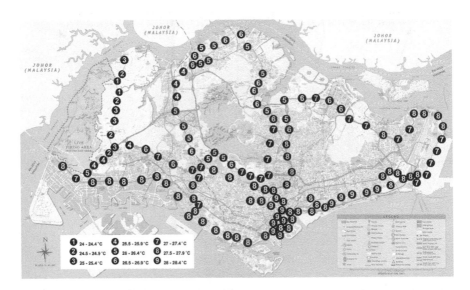

Figure 8.13 The results of the four-route survey.

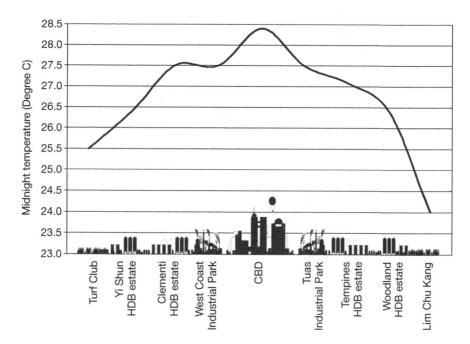

Figure 8.14 Sketch of the UHI profile in Singapore.

which can cast efficient shadow on buildings. The temperatures measured within residential areas relied on their locations. Those residential areas near to large greens normally have lower temperatures than those near to the city centre. The open space/recreation area and the forest are well planted. Therefore, they get the lowest air temperatures compared with other areas.

To further uncover the UHI effect island-wide, the ambient temperatures were analysed according to different regions. Singapore Island has been administratively divided into five regions: west region, central region, north region, north-east region and east region (see Figure 8.15). The north-western open space/recreation area and the primary forest are excluded from the zoning. From the partition of rural and urban point of view, the central region, west region and east region are mostly 'urban' areas where less vegetation is planted but a high density of buildings is constructed. The north region and north-east region are mostly 'rural' areas which are near to large green areas. The statistical data collected within every region during the mobile survey are presented in Table 8.1. The mean air temperatures of 'urban' areas (central region, west region and east region) all exceeded 27 °C. The highest mean temperature of 27.64 °C was observed in the central region. However, the mean air temperatures of 'rural' areas (north region and north-east region) were lower than 27 °C. The lowest mean value of temperature, 26.41 °C, was observed in the north region.

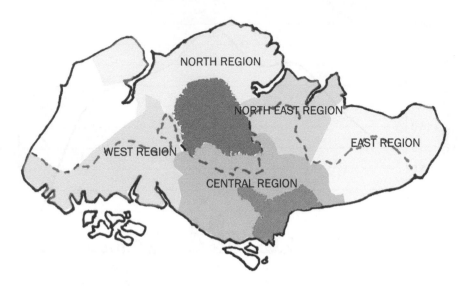

Figure 8.15 The zoning of Singapore.

Table 8.1 Statistical data for the five regions.

	Sample No.	Mean	Max	Min	Standard Deviation	Standard Error
West region	22	27.01	27.52	25.56	0.58	0.12
Central region	56	27.64	28.31	25.95	0.59	0.08
North region	17	26.38	26.73	25.56	0.39	0.10
North-east region	12	26.84	27.52	26.34	0.42	0.12
East Region	24	27.54	28.31	27.12	0.41	0.08

In order to determine whether there is a significant difference between the mean values of different regions, the Tukey-Kramer test was used. The results are given in Table 8.2. Using the 0.05 level of significance, the mean temperatures between the west and north-east regions, central and east regions, north and north-east regions respectively are not different while the rest are significantly different.

Besides the above five regions, data collected from the north-west open/ recreation area and CBD area were statistically compared (see Table 8.3). This indicates that the mean values of two areas are significantly different. The mean temperature of well-planted areas is 3.07 °C lower than that of the CBD area. The cooling effect of plants on the surrounding environment is remarkable. It is also necessary to mention that the standard deviation of data collected in the CBD area is quite small (0.22) compared with other

Table 8.2 Results of the Tukey-Kramer test (level of significance = 0.05).

	Absolute Difference	Std. Error of Difference	Critical Range	Results
West and central regions	0.632857	0.09249415	0.3570	Means are different
West and north regions	0.623824	0.11870494	0.4582	Means are different
West and north-east regions	0.168333	0.13191974	0.5092	Means are not different
West and east regions	0.536250	0.10850113	0.4188	Means are different
Central and north regions	1.256681	0.10179238	0.3929	Means are different
Central and north-east regions	0.801190	0.11693436	0.4514	Means are different
Central and east regions	0.096607	0.08968456	0.3462	Means are not different
North and north-east regions	0.455490	0.13859776	0.5350	Means are not different
North and east regions	1.160074	0.11652904	0.4498	Means are different
North-east and east regions	0.704583	0.12996527	0.5017	Means are different

Table 8.3 Statistical comparison of north-west open area and CBD area.

	Sample No.	Mean	Max	Min	Standard Deviation	Standard Error
Open area/ recreation	10	25.01	25.95	24.30	0.51	0.16
CBD area	12	28.08	28.31	27.70	0.22	0.06

regions. The possible answer may be that the 'canyon' effect within the high-density CBD area and the weak wind produced stable distribution of temperatures.

Conclusions

1   The satellite image shows a broad picture of temperature variance between the 'rural' and the 'urban' areas. This indicates the occurrence of the UHI effect during the daytime in Singapore. The 'hot' spots are normally observed on largely exposed hard surfaces in the urban context. It is understandable since the hard surfaces absorb a lot of solar radiation and incur high surface temperatures which will eventually contribute to the rising of the ambient temperatures. The satellite images also show some cool spots which are mostly observed on the surface of greenery and catchments.

2   In the historical analysis of long-term climatic data of Singapore, four meteorological stations were chosen from the National Weather Observational Network with data coverage of at least ten years. Yearly mean dry-bulb temperatures were analysed and found to be rising significantly in Changi. At the other three stations, regression results show that the yearly mean temperatures have either not changed or are warming at a much slower rate. A comparison between Changi and Tengah in terms of climatic changes and physical development over the period 1982–2001 reveals that the warming in Changi is highly likely to be attributed to an increasing UHI effect. This is because no significant warming trend is detected in the dry-bulb temperature series of Tengah and the area has not undergone any form of urbanization in the last two decades. Minimum temperature differences between the two stations show a significant increasing trend, which is found to be highly correlated to the air traffic volume at Changi Airport. This implies that the anthropogenic heat from the heavy air traffic is likely to be the major cause of the heat island effect in Changi, since heat accumulated amidst the buildings and structures during the day is freed at night, preventing night-cooling of the atmosphere. The warming trend in both the mean and minimum temperature differences not only reflects the long-term

fluctuations in the urban heat island but also shows that the UHI effect at Changi is becoming increasingly effective.

3   The results of the pilot field survey showed clear evidence of the occurrence of the UHI effect in Singapore, with the high-density high-rise commercial areas showing around 2 °C higher temperatures. Even though the dense urban fabric provides solar shading within deep street canyons, which is obvious from the lower daytime temperatures, such canyons become heat traps due to multiple solar reflection and reduced albedo.

4   The two mobile surveys successfully fulfilled the research objectives set previously. The first objective was to detect the occurrence of an urban heat island effect in Singapore. The maximum difference of 4.01 °C was observed between well planted areas and the CBD area. In addition, the mapping of temperature has shown a clear variation of temperature from southern 'urban' areas to northern 'rural' areas. All these indicate the severity of the UHI effect in Singapore. The second objective was to evaluate the cooling effect of green areas at macro level. From both mobile surveys, the survey routes near to large green areas (the north-west open space and the primary forest) experienced lower temperatures compared with other land uses like industrial areas, residential areas, the CBD area and the airport. Both lowest temperature and mean temperature, 24.3 °C and 25.01 °C respectively, were observed in a well planted area – Lim Chu Kang. On the other hand, places with less planting always have higher temperatures. The CBD region, the high-density and high-rise commercial area, has the highest temperature and mean temperature of 28.31 °C and 28.08 °C respectively. It can be concluded that large green areas definitely have a positive effect on mitigating the UHI effect in the island.

Reference

Chen, P., Liew, S. C. and Kwoh, L. K. (2001). Dependence of urban temperature elevation land cover types. *The Proceedings of 22nd Asian Conference on Remote Sensing*. Singapore International Convention and Exhibition Centre, Singapore.

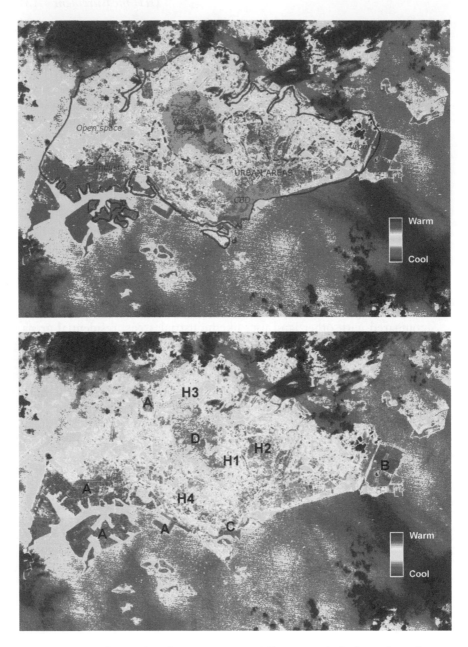

Figure 1 Coincidence of surface temperature difference with the boundary of 'rural' and 'urban' areas (top) and hot spots (bottom) in Singapore.

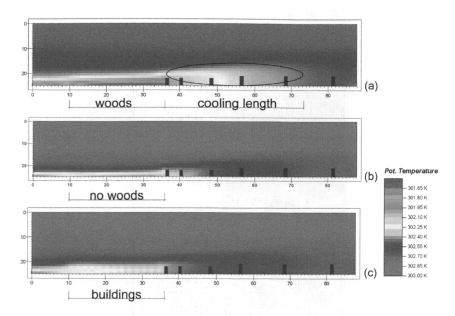

Figure 2 Comparison of section views of scenarios with woods (a), without woods (b), and with buildings replacing woods (c) at 0000 hr.

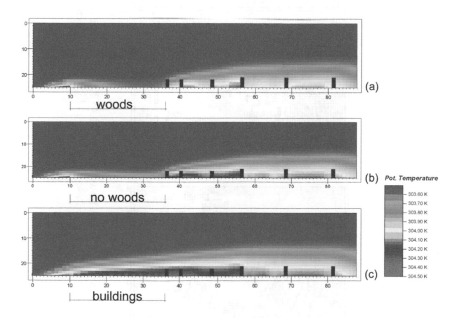

Figure 3 Comparison of section views of scenarios with woods (a), without woods (b), and with buildings replacing woods (c) at 1200 hr.

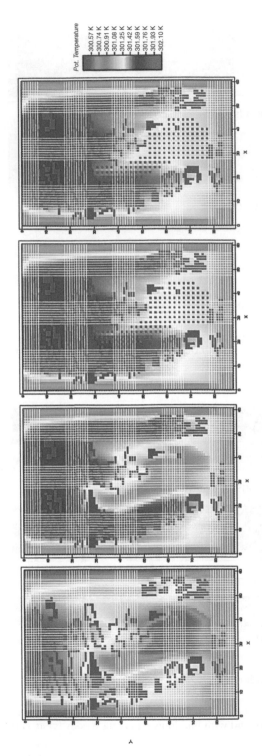

Figure 4 Temperature profile (lower limit: 303.45 K; higher limit: 301.8 K) for the different scenarios for z = 2 m at 0600 hrs (from L to R, scenario 1, scenario 2, scenario 3 with plot ratio of 1, scenario 3 with plot ratio of 3). (See also colour plates)

Pot. Temperature

300.57 K
300.74 K
300.91 K
301.08 K
301.25 K
301.42 K
301.59 K
301.76 K
301.93 K
302.10 K

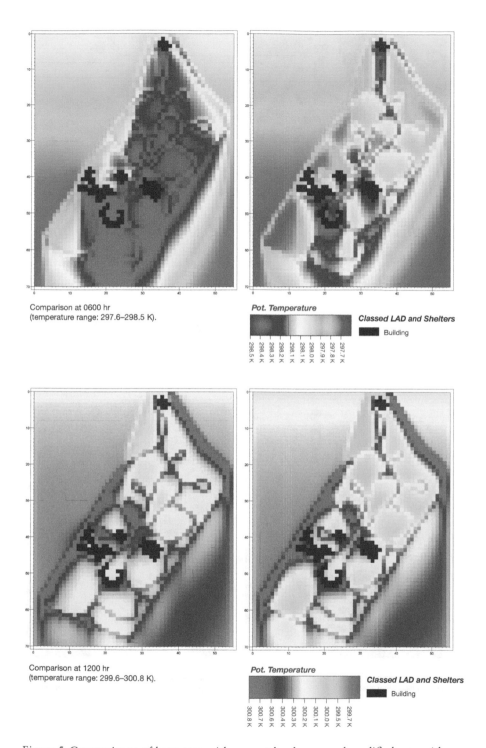

Comparison at 0600 hr
(temperature range: 297.6–298.5 K).

Pot. Temperature

Classed LAD and Shelters

Building

298.5 K 298.4 K 298.3 K 298.2 K 298.1 K 298.1 K 298.0 K 297.9 K 297.8 K 297.7 K

Comparison at 1200 hr
(temperature range: 299.6–300.8 K).

Pot. Temperature

Classed LAD and Shelters

Building

300.8 K 300.7 K 300.6 K 300.4 K 300.3 K 300.2 K 300.1 K 300.0 K 299.5 K 299.7 K

Figure 5 Comparisons of base case with current landscape and modified case with
very extensive distribution of trees in CBP.

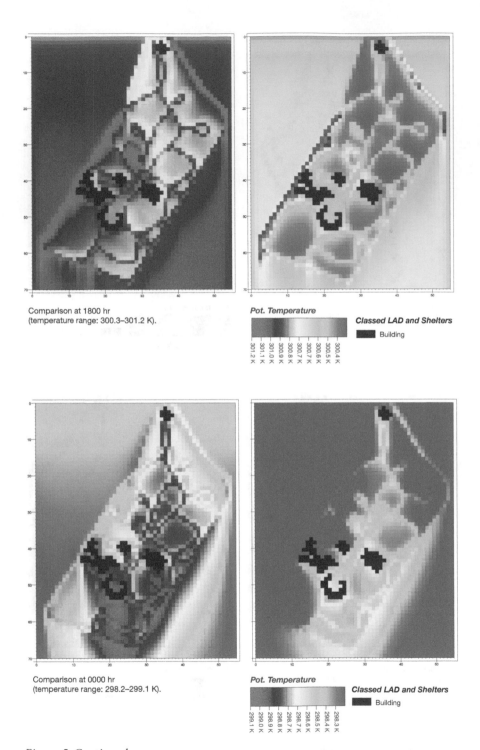

Comparison at 1800 hr
(temperature range: 300.3–301.2 K).

Pot. Temperature

301.2 K 301.1 K 301.0 K 300.9 K 300.8 K 300.7 K 300.7 K 300.6 K 300.5 K 300.4 K

Classed LAD and Shelters
■ Building

Comparison at 0000 hr
(temperature range: 298.2–299.1 K).

Pot. Temperature

299.1 K 299.0 K 298.9 K 298.8 K 298.7 K 298.7 K 298.6 K 298.5 K 298.4 K 298.3 K

Classed LAD and Shelters
■ Building

Figure 5 Continued

Figure 6 Comparison of G1 and G3 (1 April 2004).

Figure 7 Comparison of G1 and G3 (3 November 2004).

Figure 8 Comparison of G2 and G4 (1 April 2004).

Figure 9 Comparison of G2 and G4 (3 November 2004).

Figure 10 Infrared pictures of the metal roof with greenery on 5 December 2005.

Figure 11 Cross-comparison of the planted roof and the exposed roof.

Case study II
Urban parks

Preserving green areas in a built environment is one of the primary strategies for mitigating the UHI effect. Green areas play significant roles in regulating the urban climate and creating an oasis effect at macro level. In Singapore, rapid population influx has led to demands for converting natural areas to pubic housing. The replacement should be carefully evaluated before any valuable benefits of green areas are lost.

Introduction

In order to explore the thermal impacts of large city greens on the surroundings in the tropical climate, two field measurements were carried out in Bukit Batok Nature Park (BBNP) and Clementi Woods Park (CWP) in Singapore. Some information related to the two parks and the measuring points are listed in Table 8.4 and shown in Figure 8.16. In BBNP, five measuring points were selected within the park while another five points were chosen in-between surrounding residential blocks. All measuring points were lined up with an interval of around 100 metres. In CWP, six measuring points were placed evenly within the narrow strip of the park while the rest of the points were arranged in-between HDB blocks and Kent Vale. Besides using Hobo data logger, Leaf Area Index (LAI) analyser and weather station were also used to measure LAIs and weather data in CWP. In order to explore the possible energy savings near the parks and the impact of change of land use, some simulations were carried out based on the data derived from the measurements.

Table 8.4 Bukit Batok Nature Park and Clementi Woods Park.

	Size	*Location*
Bukit Batok Nature Park (BBNP)	36 ha	Located in the centre of Singapore
Clementi Woods Park (CWP)	12 ha	Located in the western part of Singapore

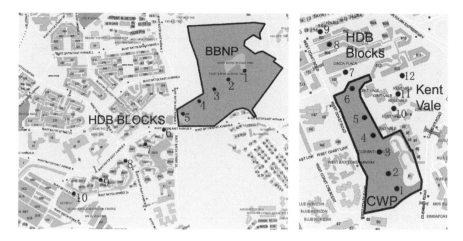

Figure 8.16 Measuring points around the two parks (left: BBNP; right: CWP).

Discussion and observations

BBNP

To explore the cooling effect of BBNP, the average temperatures obtained at different locations were compared (see Figure 8.17). Within BBNP, it could be found that most average temperatures were relatively lower than those measured in HDB blocks. From locations 1 to 4, the average temperatures ranged from 25.2 to 25.5 °C. For location 5, the average temperature was slightly higher since it is located at the edge of BBNP. Furthermore, the location is near to the car park and highway. The anthropogenic heat generated by parked vehicles has probably influenced the readings. There is an orderly elevation of average temperatures for locations within the surrounding HDB blocks. This indicates that the park has a cooling impact on the surroundings but it is limited by the distance. The highest average temperature was observed at location 9. It was 1.3 °C higher than the average temperature obtained at location 6 which is the nearest HDB location to BBNP. Location 10 had lower average temperatures compared with location 9. It was because the location is at the edge of the dense HDB neighbourhood. The impact from buildings on location 10 may not be as much as that on locations within the blocks. Another interesting difference between the park and surrounding HDB blocks is their standard deviation (SD) (see Figure 8.18). The SDs obtained in the park ranged from 1.8 to 2.1 (locations 1 to 4) while the SDs in the built environment ranged from 2.0 to 3.2 (locations 5 to 10). Basically, the higher the average temperatures (which are mostly observed in the built environment), the higher the SD. This indicates that a well planted location, compared with a built environment,

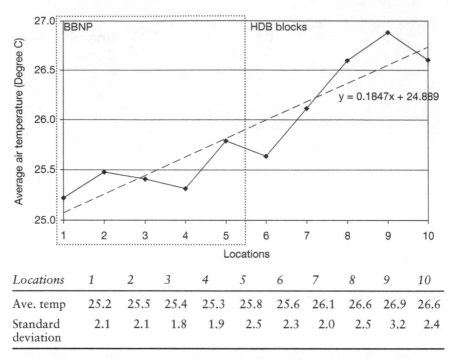

Locations	1	2	3	4	5	6	7	8	9	10
Ave. temp	25.2	25.5	25.4	25.3	25.8	25.6	26.1	26.6	26.9	26.6
Standard deviation	2.1	2.1	1.8	1.9	2.5	2.3	2.0	2.5	3.2	2.4

Figure 8.17 Comparison of average air temperatures obtained at different locations in BBNP during the period from 11 January to 5 February 2003.

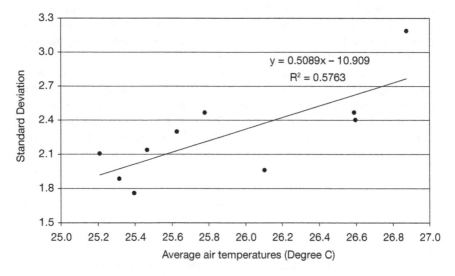

Figure 8.18 Correlation analysis between standard deviation and average temperature for every measuring location.

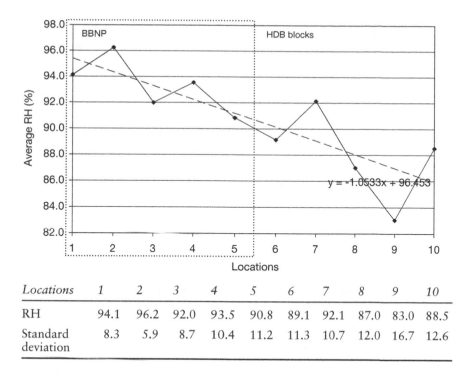

Locations	1	2	3	4	5	6	7	8	9	10
RH	94.1	96.2	92.0	93.5	90.8	89.1	92.1	87.0	83.0	88.5
Standard deviation	8.3	5.9	8.7	10.4	11.2	11.3	10.7	12.0	16.7	12.6

Figure 8.19 Comparison of average relative humidity obtained at different locations in BBNP during the period from 11 January to 5 February 2003.

may have better ability to stabilize the fluctuation of ambient air temperature. Figure 8.19 shows the comparison of average relative humidity obtained at different locations. Inversely, average relative humidity obtained in BBNP is higher than that obtained from surrounding HDB flats. All average relative humidity obtained from BBNK was over 90 per cent. This indicates the humid environment in the park.

To further explore the temperature profiles, a clear day was selected (see Figure 8.20). Basically, the findings accord with the average temperature profiles obtained during the longer period. The interesting finding is that both the temperatures measured within the park and in the residential blocks were higher than those averaged over a period of time. This indicates that a clear day with more incident radiation can cause high temperatures everywhere. However, the lowest temperature obtained within the park increased by less than 2 °C whereas the highest temperature derived near the buildings increased by more than 2.5 °C. This finding supports the view that a green area has better climatic control compared with a built environment.

The measurement in BBNP was carried out over a period of 26 days. The lower temperatures were observed within BBNP where there are dense trees.

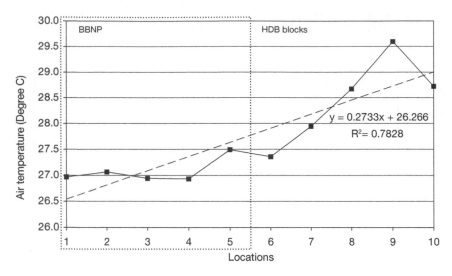

Figure 8.20 Comparison of average ambient air temperatures measured within BBNP on a clear day (23 January 2003).

They can be defined as a 'cooling source' in this case. To find out the correlation between a cooling source and locations away from it, a correlation analysis was done (see Figure 8.21). Location 3 has been defined as a benchmark since it had lower average temperatures and standard deviations over a period of 26 days. The elevation of ambient temperatures can be observed at locations in terms of the gradient. Locations 6 and 7 have relatively smaller gradients. The temperature difference between them and location 3 was not much. But for locations 8, 9 and 10, the gradients become larger. Location 9 has the largest gradient and the highest temperature difference between location 9 and location 3 was experienced. Figure 8.21 presents the correlation analysis among the benchmark, location 6 (just next to the park), and location 9 (the worse case). The rest of the scenarios are in-between the two extreme cases.

A typical eight-storey commercial building was modelled in a simulation. The cooling load of the commercial building was consecutively calculated when it was placed inside the park, 100 m, 200 m, 300 m and 400 m away from the park. The results are given in Table 8.5. It can be observed that the impact of the park was reflected by the clear difference among the cooling loads. The lowest one, 9077 kWh, was observed when the building was placed inside the park while the highest one, 10,123 kWh, was recorded when the commercial building was built 400 m away from the park (location 9 in the field measurement). The energy consumptions of the rest of the locations are within the range defined by the above two locations. This accords with the average temperature profile observed in the field measurement in

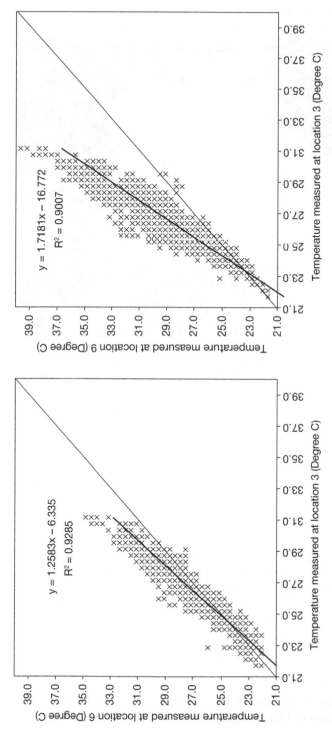

Figure 8.21 Correlation analysis of the benchmark (location 3) in the park and two locations away from the park (11 January to 5 February 2003).

Table 8.5 Comparison of cooling loads for different locations.

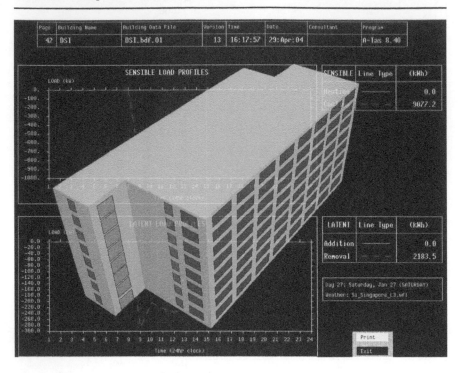

	cooling load (kkWh)	Energy savings compared (with the scenario of 400 m)
In the park	9077	10%
100 m from the park	9219	9%
200 m from the park	9383	7%
300 m from the park	9672	4%
400 m from the park	10123	0%

BBNP. It may not be realistic to hope that a commercial building can be built inside a park but it is highly possible that the building can be built near to a park or greenery. The cooling loads analysis provides evidence that buildings can benefit from a nearby park. For example, 9 per cent cooling energy could be saved if an eight-storey commercial building was built closer to the park (moving from 400 m away to 100 m away).

CWP

Comparisons of temperatures averaged over a period from 16 June to 1 July 2003 in CWP are given in Figure 8.22 and Figure 8.23. It can be observed that the lowest average temperature, 25.7 °C, was experienced at location 1 while the average temperatures ranged from 27.2 to 27.5 °C at the rest of the locations within CWP. The difference between location 1 and the rest of the locations in CWP is from 1.5 to 1.8 °C. This can be explained by the arrangement of plants within CWP through its Leaf Area Index (LAI) values (see Table 8.6). The southern part (around location 1) of CWP is a strip of very dense forest where the LAI readings measured were 7.11 and 7.23. Both shading and evaporative cooling are outstanding in the area. The rest of CWP was planted with trees and their density, especially measured on the ground, is not so encouraging. The measured LAIs ranged from 2.21 to 4.02. On the other hand, the ambient air temperatures measured within the park were relatively lower than those obtained within the buildings. The elevation of average air temperatures can be observed in both HDB blocks and Kent Vale blocks. This also follows the rule that the further away from the park the higher the average temperatures. Similar to the observations in BBNP, the standard deviations of ambient temperatures are smaller for locations within the park compared with those obtained from locations within the built environment.

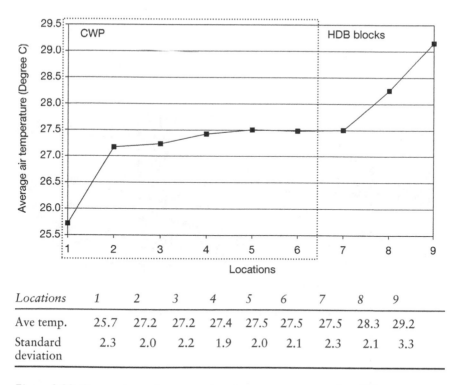

Locations	1	2	3	4	5	6	7	8	9
Ave temp.	25.7	27.2	27.2	27.4	27.5	27.5	27.5	28.3	29.2
Standard deviation	2.3	2.0	2.2	1.9	2.0	2.1	2.3	2.1	3.3

Figure 8.22 Comparison of average air temperatures measured at different locations in CWP and HDB blocks (16 June to 1 July 2003).

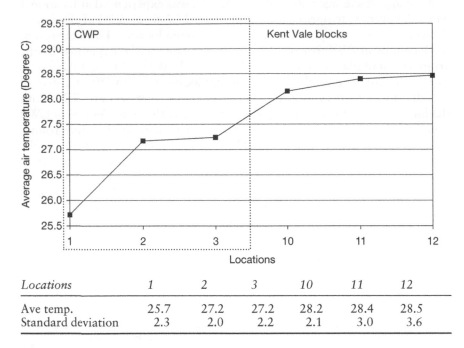

Locations	1	2	3	10	11	12
Ave temp.	25.7	27.2	27.2	28.2	28.4	28.5
Standard deviation	2.3	2.0	2.2	2.1	3.0	3.6

Figure 8.23 Comparison of average air temperatures measured at different locations in CWP and Kent Vale apartment (16 June to 1 July 2003).

Location 1: Dense part   Location 3: Sparse part

Figure 8.24 Location where the LAI was measured in the park.

Table 8.6 The LAI measurement in CWP.

Location	LAI readings
1	7.23
2	7.11
3	2.32
4	4.02
5	2.21
6	2.43
7	2.29

Besides the comparison of ambient air temperatures, the correlation analysis between solar radiation and air temperatures at different locations has also been done (see Figure 8.25). The solar radiation data was taken from the weather station located near the CWP. On a clear day (19 June 2003), only a period of time from 0800 to 1340 hr when the solar radiation continuously increased to its peak level was used to do the analysis. Basically, temperatures will increase with the elevation of solar radiation at all locations. The trend line of location 1 is at the bottom of the profile followed by those derived within CWP and the nearby area. The trend lines obtained from the locations further away from the park, such as locations 9, 11 and 13, are at the top of the profile. All these accord with the previous analysis, but the interesting phenomenon is that the gradients of these trend lines are more or less the same. In other words, the increase of temperature due to the

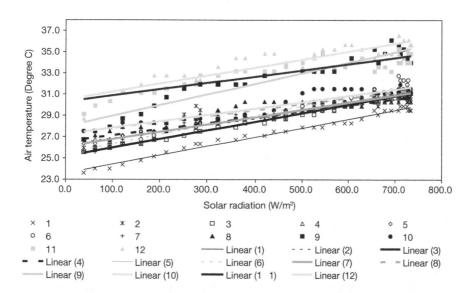

Figure 8.25 Correlation analysis between solar radiation and air temperatures measured at all locations on a clear day.

elevation of solar radiation is more or less the same for all locations. The evaporative cooling impact of plants on ambient air seems to be blurred in this manner. Probably the shading effect of trees dominates when solar radiation is in the increasing phase towards its peak level.

Envi-met was employed to compare the thermal conditions with and without the park in the area. The initial boundary parameters were set according to the real site condition. Based on the preliminary analyses of weather data obtained from the field measurement, a clear sunny day was chosen. The setting of the air temperature, wind speed and relative humidity were based on the mean values of the weather data. The wind direction was set from south to north. Three scenarios were created as follows:

- Scenario 1 – Original woods
- Scenario 2 – Removal of woods
- Scenario 3 – Replacing woods with buildings (plot ratio of 1 and 3 respectively).

Four typical time scenarios, 0000 hr, 0600 hr, 1200 hr and 1800 hr, were selected for the analysis. Figure 8.26 (see also colour plates) shows the comparison observed at 0000 hr. It is obvious that a planted area can create a low-temperature zone in its leeward area. The length of the area is almost similar to the length of the green area. The simulation results show that when the points are closer to the green area, lower temperatures were observed. The maximum height of the low-temperature zone is around 70 to 80 metres. In this simulation model, the height of the tallest buildings (HDB block) is 66 m. Therefore, we can conclude that both the upper and lower parts of tall buildings can get a cooling benefit from vegetation at night if they are located near a green area. When the vegetation is totally removed from the woods, only a very small patch of low temperature is found in the leeward area. When plants are totally replaced with buildings, the low-temperature zone is replaced by a high-temperature zone.

Figure 8.27 (see also colour plates) illustrates the daytime scenarios of the simulation. This is totally different from the night-time ones in terms of vertical temperature distribution. The low-temperature zone in the leeward area of the greenery lost its clear boundary compared with the one observed at night. This can be explained by the fact that the built area is exposed to the insolation during daytime and incurs high ambient temperature near the hard ground. Compared with the high temperature in the built environment, the cooling effect of the woods does not prevail during the daytime. However, lower temperature still can be observed in the leeward area of the woods compared with the scenarios without woods or replacing woods with buildings.

For the cross-comparison of the effects of different interventions, the model results are presented side-by-side to be evaluated qualitatively. Figure 8.28 (see also colour plates) shows the temperature profile of the four scenarios

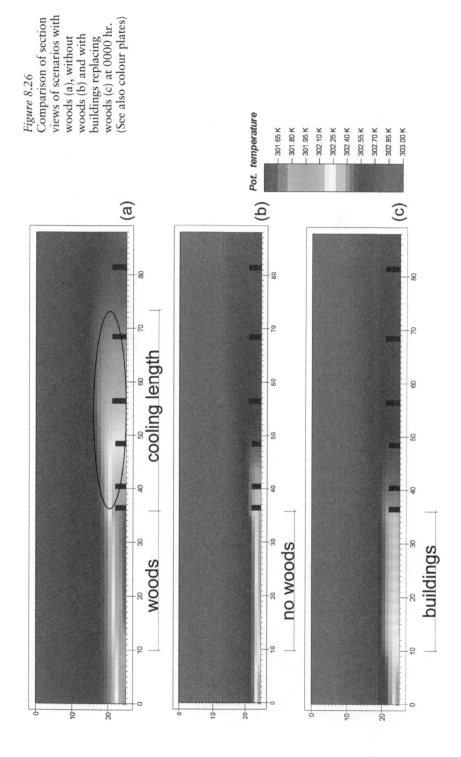

Figure 8.26
Comparison of section views of scenarios with woods (a), without woods (b) and with buildings replacing woods (c) at 0000 hr. (See also colour plates)

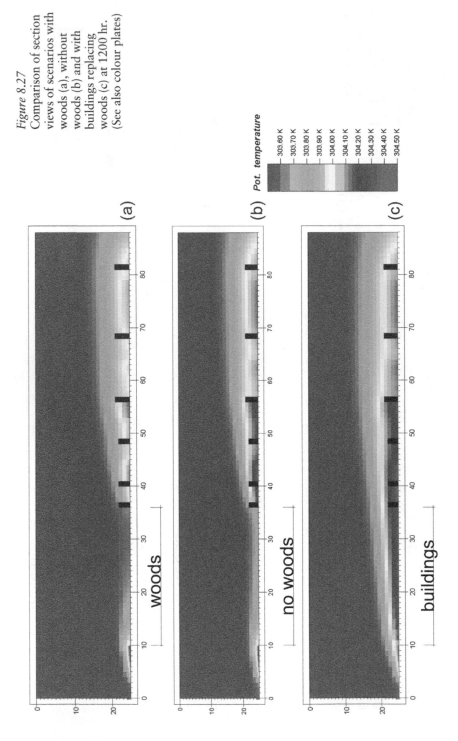

Figure 8.27
Comparison of section views of scenarios with woods (a), without woods (b) and with buildings replacing woods (c) at 1200 hr. (See also colour plates)

Pot. *temperature*

303.60 K
303.70 K
303.80 K
303.90 K
304.00 K
304.10 K
304.20 K
304.30 K
304.40 K
304.50 K

(a)

woods

(b)

no woods

(c)

buildings

at 0600 hrs. In scenario 1, it is observed that the coolest region is in Clementi Woods itself due to the greenery at about 300 K. The cooling effect of Clementi Woods can be seen in Kent Vale and the lower portion of Ginza Plaza at around 301.1 K. In HDB areas, the effect is less pronounced at 301.5 K.

When the vegetation is removed in scenario 2, the temperature in the woods area has been elevated to about 301 K. However, the moisture level in the soil does not cause the temperature to be similar to those on the hard pavement areas. Subsequently, the cooling effect in Kent Vale and lower Ginza Plaza is reduced to a certain extent. In the HDB block areas, the temperature has risen to about 301.8 K. It is to be noted that the model assumes that the source of water in the soil is non-depleting. In reality, however, it is expected that the water will dry up after some time and thus the cooling effect will be minimal.

When the vegetation is replaced with hard pavement and buildings, it can be seen that the whole area now has a higher temperature at about 301.5 K. There appears to be a loss of cooling effect on Kent Vale and lower Ginza Plaza where the temperature is about 301.8 K. Similarly, the areas in the HDB blocks have the highest temperature of 302.2 K. Comparing the scenarios of buildings with plot ratios of 1 and 3, it can be seen that in the latter, the temperature profile is modestly better than in that of the former. This is attributed to the higher wind velocities through the taller buildings in scenario 3 with the plot ratio of 3. The taller buildings create higher negative pressure causing higher velocities of air to move through the sides of the building thus increasing ventilation and reducing temperature.

Besides the air temperature, wind speed and humidity were also examined in the scenarios. The cooling effect of greenery can be confirmed by the simulation. It has been found that the cooling effects of the greenery area on surrounding areas are strongly related to distance to the greenery area, wind direction and building layout. The qualitative analysis of the temperature data showed that the coolest region was in the Clementi Woods due to the existence of the green area. This is confirmed by quantitative analysis which showed that Clementi Woods is 0.3 to 0.6 °C lower than other zones. Also, the temperature difference between Clementi Woods and surrounding areas at night-time is higher than that in the daytime. The relative air temperature in Clementi Woods varies from 0.15 to 0.74 °C. In the cross-comparison of the four scenarios for temperature, the best cooling effect on the surrounding built-up area is observed in the base case scenario (with vegetation). This effect is reduced when the vegetation is removed leaving behind the soil and the cooling effects are drastically reduced when buildings are erected. The reduction of the air temperature in Clementi Woods can reach 0.2 to 0.5 °C. Specific humidity levels are highest in regions where the temperature is highest. This is due to the fact that at elevated temperatures, the air has the capacity to hold more moisture. Consequently, specific humidity is higher in the daytime than that at night-time with levels varying from 12.9 to 18.1 g/kg.

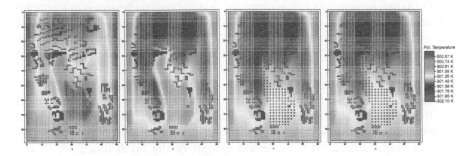

Figure 8.28 Temperature profile (lower limit: 303.45 K; higher limit: 301.8 K)
for the different scenarios for z = 2 m at 0600 hrs (from L to R,
scenario 1, scenario 2, scenario 3 with plot ratio of 1, scenario 3 with
plot ratio of 3). (See also colour plates)

Wind velocity profiles showed that the highest values are for regions with no
obstruction found along the roads and open areas. The wind velocity over
Clementi Woods is seen to be reduced by the vegetation. This pattern is
consistent across the day. In the cross-comparison of the four scenarios for
wind velocity, the wind velocity became high when the vegetation was
removed in scenario 2. But this effect is restricted to the lower part of the
HDB only. Wind velocity is relatively higher between blocks with higher plot
ratios. Also, scenario 2 has the highest speed due to no blockage of vegetation. Higher wind speed zones appear in the street canyon due to the channel
effect from the two sides of HDB blocks. The channel effect in the wooded
area of the scenario with a plot ratio of 3 can raise the wind speed and
produce better ventilation.

Conclusion

The cooling impact of parks is reflected through not only the lower
temperatures in the parks but also the lower temperatures in the nearby built
environment (see Table 8.7). In general, the average temperatures obtained
in both parks were lower than those in the surrounding environment. For
the surrounding built environment, the closer to the park, the lower the
temperature experienced. Maximally, 1.3 °C difference in average temperature was observed at locations around the parks. The temperature difference
was caused by green areas and this may lead to savings of cooling energy and
thermal comfort for residents. Another interesting phenomenon is that the
range of the standard deviation obtained from the planted area is smaller
than that derived from the built environment. This provides evidence that
the ability of vegetation to stabilize the fluctuation of the temperatures is
greater than that of building materials. The temperatures measured within
the parks also have a strong relationship with the density of plants since

Table 8.7 Temperature and standard deviation derived from the measurements.

	In the parks	In surrounding built environment
Average temp. variation (Degree C)	25.2 – 27.5	25.6 – 29.2
Standard deviation variation	1.8 – 2.3	2.0 – 3.6

plants with higher LAIs may cause lower ambient temperatures. The cooling range incurred by large green areas may be related to the wind direction, layout and height of surrounding buildings, and the foliage density (LAIs) of parks.

The temperature difference was caused by the vegetation area and it may result in savings of cooling energy in surrounding buildings. Results derived from the energy simulation support the observations of field measurements. Energy may be saved when buildings are built near to parks. Maximally a 10 per cent reduction of the cooling load was observed in the simulation. Two things should be noted. To build a structure near to a park does not mean it should be built inside a park to acquire the best performance. The energy used for dehumidifying is not considered in the simulation although it may not be significant in buildings with good ventilation. Normally, the nearer to the park, the higher the relative humidity to be encountered.

The Envi-met simulation also supported the data generated from the field measurement. It indicates that a park has a significant cooling effect on surroundings during both day and night. Moreover, the Envi-met simulation also illustrates the truth that the loss of greenery may cause bad thermal conditions not only in the original park area but also in the surroundings, especially when greenery is replaced by buildings/hard surfaces.

In summary, the importance of big city greens has been confirmed through measurements and simulations under the tropical climate. But the benefits of parks are not limited to the thermal aspect only. They are also invaluable from the ecological and social points of view. More concern should be paid to the preservation of green areas in cities rather than simply replacing them with buildings.

Case study III
Trees in housing developments and industrial areas

Singapore's rapid urbanization and industrial growth have exerted a heavy toll on its flora and fauna. Habitats were destroyed when coastal areas were reclaimed to provide for more land, and forests were cut down to provide residential and commercial sites. Large green areas become a luxury in these circumstances. Concept Plan 2001 and Master Plan 2003 proposed by the Urban Redevelopment Authority (URA) expect to set aside land to maintain the current standards of park and vegetation, including parks, open spaces, interim greens, park connectors, promenades, nature reserves, nature areas and waterbodies. It is believed that trees planted within housing developments and along roadsides not only have an ornamental function but also a role in regulating the climate.

Punggol site versus Seng Kang site (housing development)

Introduction

Two residential sites were chosen for the measurement, the Punggol site and the Seng Kang site. The Punggol site is a developed, residential site with moderate vegetation on ground level and a rooftop garden. It consists of Block numbers 127, 128A, 128B, 128C, 128D and 128 (carpark) situated at Punggol Field Walk Road. The Seng Kang site is also a developed residential site but with considerable less vegetation at the moment and no vegetation on the top of the carpark. It consists of Block numbers 183, 183A, 183B and 184 (carpark) situated at Cres Rivervale Road. Calculations have shown that the Punggol site has vegetation coverage roughly three times more than the Seng Kang site (see Table 8.8). The measurement was conducted over a two-week period from 21 September to 5 October 2003. Eight measuring points were evenly spread out in each site.

Discussion and observations

Figure 8.29 presents a comparison of the temperatures measured at the two sites over a period of two weeks. Site 1 has a slightly lower average

Table 8.8 Green coverage at two sites

	Area covered (m²)	Area covered/ total area surveyed (m²)	Percentage (%)
a. Punggol site			
Grass, gravel/soil (<0.15 m)			
Shrubs (0.15m–1.5m)	2383	2383/10719	22
Other vegetation (>1.5 m)			
Pavement/Building	8336	8336/10719	78
b. Seng Kang site			
Grass, gravel/soil (<0.15 m)			
Shrubs (0.15m–1.5m)	720	720/10719	7
Other vegetation (>1.5 m)			
Pavement/Building	9999	9999/10719	93

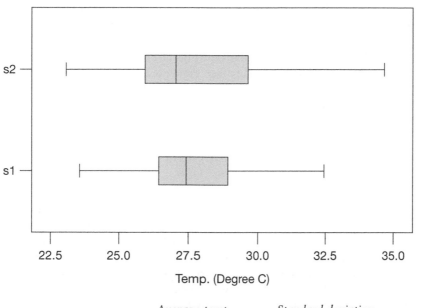

	Average temp.	Standard deviation
Site 1: Punggol site	27.6	1.72
Site 2: Seng Kang site	27.7	2.44

Figure 8.29 Comparison of temperatures measured at the two sites over a period of two weeks.

temperature compared with site 2. A clear difference of the maximum temperatures, about 2.3 °C, can be observed between the two sites. The higher maximum temperature was experienced during the daytime when the residential area with less greenery was easily heated up. On the other hand, site 1 has a slightly higher minimum temperature. This is probably due to the blockage of wind flow by the landscape at night. Overall, the ability of greenery in terms of climatic control can be observed from the smaller deviation in site 1.

In order to observe the performance of greenery on a clear day, the comparison of maximum, minimum and average temperatures between two sites is shown in Figure 8.30. For a short period in the morning till around 0900 hr, both sites had shown quite similar average ambient air temperature readings. During the daytime, from 0900 to 1900 hr, the temperatures measured at the Seng Kang site had consistently higher temperatures than those at the Punggol site. The temperature in Seng Kang also rose more rapidly than at Punggol and reached the peak average air temperature at around 1600 hr. The maximum average temperature difference between the two sites was 2.32 °C. Since the two sites are near to each other and have similar layouts, the difference is mainly due to the existence of greenery in Punggol. The temperatures measured in Seng Kang started to decrease at around 1700 hr and reached similar readings at Punggol at around 1900 hr. The comparison of relative humidity (see Figure 8.31) was by and large lower in Seng Gang as compared to Punggol, where the plants release moisture into the surroundings through the process of evapo-transpiration. It was over 70 per cent on average even during the daytime at the Punggol site.

Changi Business Park (CBP) measurement

Introduction

CBP is a modern business park with five multi-storey commercial buildings constructed within the boundary. The distribution of the landscape within the boundary is currently uneven. The southern part is crowded with buildings and sparse landscape (mainly turfing) while the northern tip is still vacant with dense greenery. There were altogether six measuring points evenly chosen in the southern part while one point was in the vacant area. In addition, two extra points were installed in the nearby traditional industrial area as a reference. The measurement was carried out over a period of 57 days. Envi-met simulation was carried out subsequently in order to study the impact of greenery arrangement in CBP.

Discussion and observations

Figure 8.32 presents the average temperatures calculated at different locations over a period of 21 days. It is obvious that the average temperature obtained from the vacant area with relatively extensive greenery nearby was

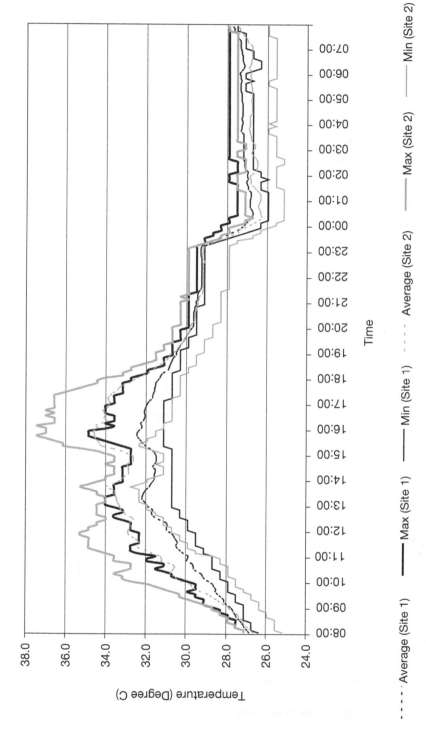

Figure 8.30 Comparison of temperatures between two sites (Site 1: Punggol site; Site 2: Seng Kang site).

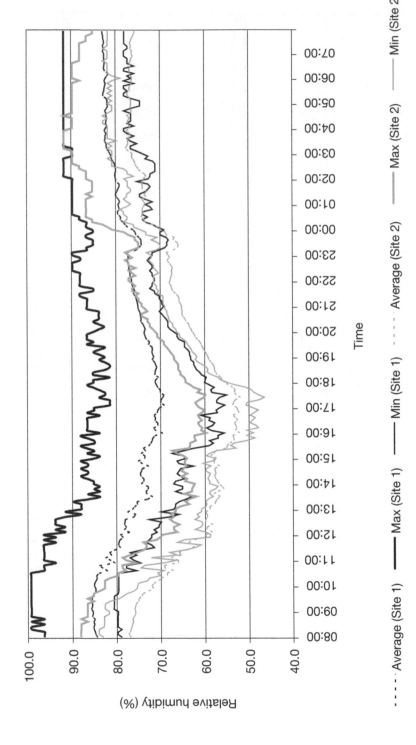

Figure 8.31 Comparison of relative humidity between two sites (Site 1: Punggol site; Site 2: Seng Kang site).

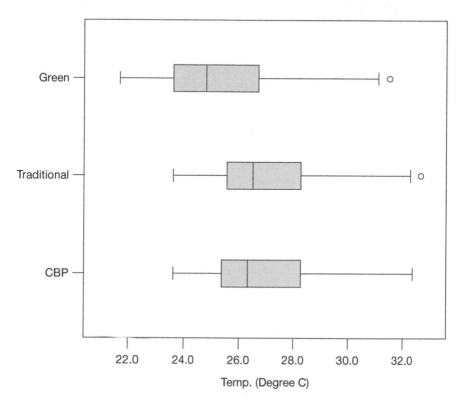

Figure 8.32 Comparison of average air temperatures obtained at different locations over a period of 21 days in December 2004.

	Temp.	*Standard deviation*
Vacant area with greenery	25.2	2.2
Traditional industrial area	27.0	1.9
CBP	26.8	1.9

significantly lower than the rest of the locations. It is important to note that the period includes not only clear days but also cloudy and rainy days. The cooling impact, around 1.6 to 1.8 °C lower on average, brought by greenery on the vacant land is impressive. By retaining as many trees as possible on site, the future development of CBP in the vacant area can definitely benefit from such temperature reduction. On the other hand, the difference, around 0.2 °C on average, between CBP and the traditional area is not so significant. This indicates that the current landscape in CBP, although it has large areas of turfing, has limited cooling impact in the vicinity.

The comparison was further carried out on a clear day (see Figure 8.33). The average temperature of the vacant area with extensive greenery was still consistently lower than the others throughout the day. This accords with the observation generated from the long period analysis and indicates the predominating cooling impact of extensive greenery on surroundings. The temperature difference between the vacant area and the industrial areas is big at night and relatively small during the daytime. This is understandable since the insolation can easily blur the cooling impact brought by the trees. The slight temperature difference between CBP and its surrounding traditional industrial area can be observed on a clear day although it is still not so significant.

Based on the results derived from the measurement, a simulation was carried out by the use of Envi-met. The aim was to investigate the possible difference in the business park (CBP) with existing landscape and with a dense distribution of trees. Figure 8.34 (see also colour plates) shows the

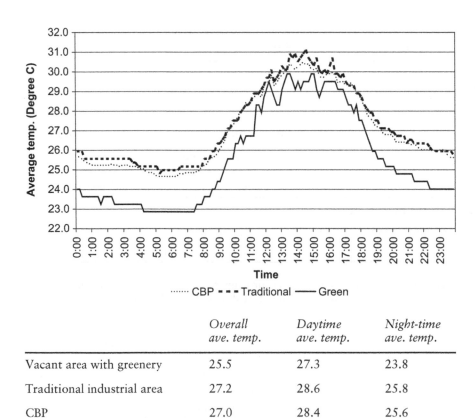

	Overall ave. temp.	Daytime ave. temp.	Night-time ave. temp.
Vacant area with greenery	25.5	27.3	23.8
Traditional industrial area	27.2	28.6	25.8
CBP	27.0	28.4	25.6

Figure 8.33 Comparison of average temperature computed every 10 mins at different locations on a clear day (20 December 2004).

visual comparison derived from the simulation. At 0600 hr, when the lowest temperature of the day should be experienced, a clear difference can be observed between the two scenarios. This reflects the cooling impact of the trees in the vicinity. At 1200 hr, the temperature difference between the two scenarios becomes unclear although greenish colour can still be observed somewhere in the modified case. This can be understood since the high solar radiation can easily blur the performance of trees by then. The very significant difference emerges at 1800 hr, around sunset. After absorbing solar heat during the daytime, the base case with more hard surfaces indeed shows pinkish everywhere. However, with the protection of the trees, the modified case shows a relatively lower temperature represented by a greenish colour. The difference extends until 0000 hr. The observations generated from the first round simulation strongly support the view that planting more trees can modify the current thermal condition in CBP. The performance of the trees can be more easily observed at night rather than in the daytime (during daytime, the impact of trees is not reflected by reducing the ambient air temperatures but by the surface temperature of the hard surfaces shaded by plants). This accords with the observations obtained from the field measurements.

Roadside trees in Tuas

Introduction

The impact of roadside trees in an industrial area was explored in three selected streets in Tuas. Tuas Avenue 2 and Tuas Avenue 8 are two parallel streets which are separated from the Raffles Golf Course by the highway. Tuas Avenue 2 has very densely planted mature trees along the roadside. Trees planted along Tuas Avenue 8 are not so mature and densely planted compared to Tuas Avenue 2. On the other hand, Tuas Avenue 2 and Tuas Avenue 8 are all streets with busy traffic during the daytime. Tuas South Street 3 is further away from Tuas Avenues 2 and 8 down to the southern part of Tuas. Compared with Tuas Avenues 2 and 8, Tuas South Street 3 is a street with very young trees and relatively light traffic. There were altogether six measuring points set up in Tuas Avenue 2 and Tuas Avenue 8 and five ones in Tuas South Street 3. All measuring points were evenly distributed in the three streets with an interval of roughly 100 m. The measurements were conducted in the Tuas area from 18 March to 15 April 2005. A simulation was carried out for exploring the possible energy savings caused by roadside trees according to the results derived from the field measurements.

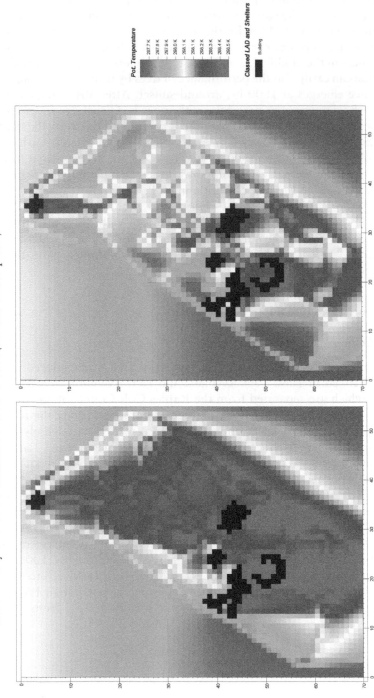

Figure 8.34 Comparison at 0600 hr (temperature range: 297.6–298.5 K). Comparisons of base case with current landscape and modified case with very extensive distribution of trees in CBP. (See also colour plates)

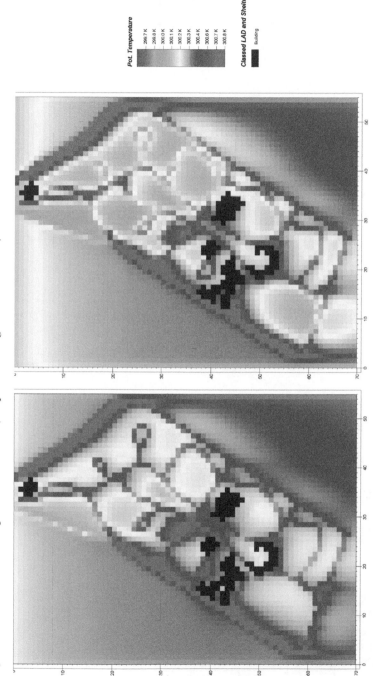

Figure 8.34 Continued. Comparison at 1200 hr (temperature range: 299.6–300.8 K).

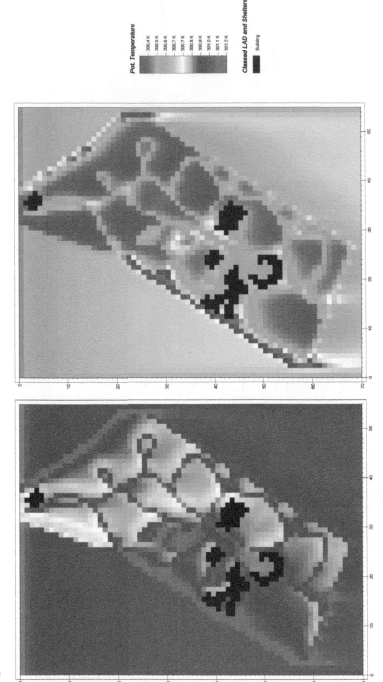

Figure 8.34 Continued. Comparison at 1800 hr (temperature range: 300.3–301.2 K).

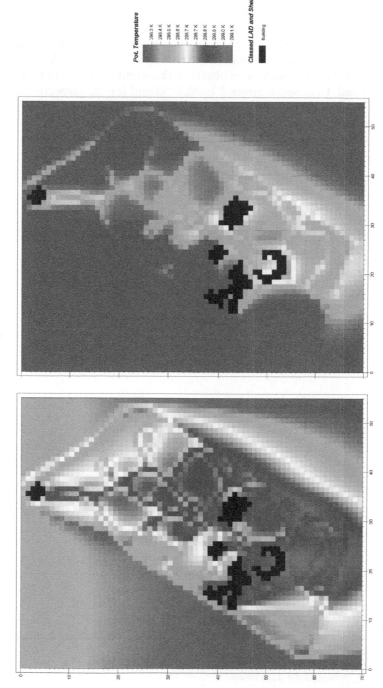

Figure 8.34 Continued. Comparison at 0000 hr (temperature range: 298.2–299.1 K).

Discussion and observations

A temperature comparison between the three selected streets in the Tuas area is shown in Figure 8.35. It is interesting to note that the mean and the median values obtained from Tuas Avenue 2, Tuas Avenue 8 and Tuas South Street 3 are in a sequence according to the density of the planted roadside trees. Tuas Avenue 2 has many mature trees with big crowns. As expected, the lowest mean temperature was observed there and it was lower than Tuas Avenue 8 and Tuas South Street 3 by 0.5 °C and 0.6 °C respectively. The mean or median temperature obtained from Tuas Avenue 8 was slightly lower than that observed in South Street 3 although Tuas Avenue 8 has much better green coverage. A possible reason for this may be interference from

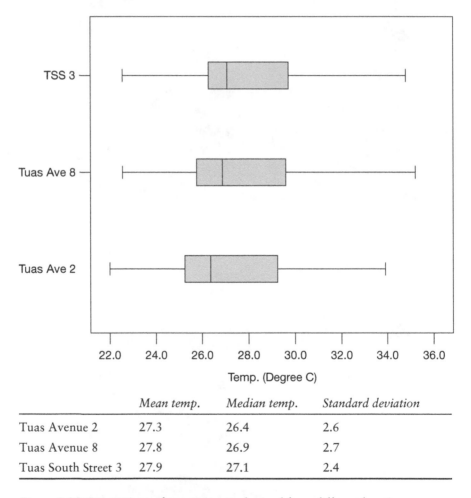

	Mean temp.	*Median temp.*	*Standard deviation*
Tuas Avenue 2	27.3	26.4	2.6
Tuas Avenue 8	27.8	26.9	2.7
Tuas South Street 3	27.9	27.1	2.4

Figure 8.35 Comparison of temperatures obtained from different locations over a period of 25 days in 2005.

the heavy traffic. The right whisker of Tuas Avenue 8 in the figure is at the highest level which indicates that high temperature was experienced due to the combined effect of the anthropogenic heat from the busy traffic and incident solar heat. On the other hand, although similar traffic conditions were experienced in Tuas Avenue 2, a lower right whisker is observed due to the climatic modification of the mature trees.

Since the average temperature difference is not so significant among the three streets over the period, the extreme condition on a clear day, 10 April 2005, was examined (see Figure 8.36). It seems that night-time was ideal for the comparison since there was less influence from the traffic. A clear sequence can be observed among the three streets. Tuas Avenue 2, as expected, had the lowest average temperature followed by those in Tuas Avenue 8 and Tuas South Street 3. A maximum difference of more than 1 °C can be observed between Tuas Avenue 2 and Tuas South Street 3. This accords with the previous observations. During the daytime, especially from 1000 to 1730 hr, the influence from the heavy traffic seems to be serious in Tuas Avenue 2 and Tuas Avenue 8. This is reflected in the higher temperatures observed on the two streets.

In order to explore the possible energy savings caused by roadside trees on a standard single-storey factory, an energy simulation was carried out (see Figure 8.37). The factory model was put into four scenarios with reference to the corresponding field measurements' results. Case 1 is the environment of Tuas South Street 3 where only young trees without effective tree shading is observed. Case 2 is the environment of Tuas Avenue 8 where medium density of road trees and very heavy traffic during the daytime is experienced.

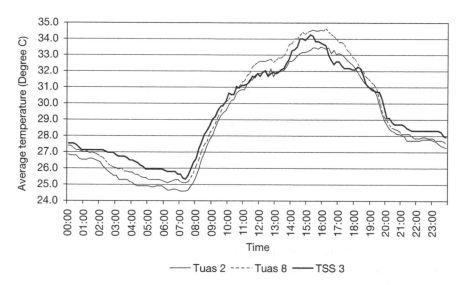

Figure 8.36 Comparison of average temperatures measured in the three streets on a clear day.

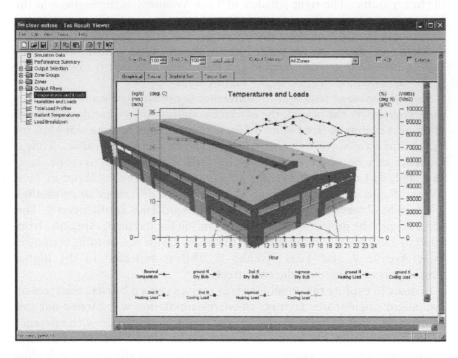

Figure 8.37 The stand-alone factory model and TAS simulation.

Case 3 and case 4 are all derived from the condition of Tuas Avenue 2. The difference is that case 3 employed the averaged temperatures over the whole street while case 4 used the lowest temperature profile observed among all the locations in the street. Figure 8.38 shows the result of the simulation. In general, it is clear that cooling energy savings can be achieved when the stand-alone factory is placed into an environment with mature roadside trees. By using case 1 as the benchmark (see Table 8.9), it can be easily observed that 5 per cent of energy can be saved in an environment with fairly dense roadside trees and up to 23 per cent of energy saving can be achieved in an environment with extremely dense roadside trees on a clear day. Roadside trees can also contribute to energy savings, ranging from 11 per cent to 38 per cent, on a cloudy day. For case 2, the potential saving is almost zero. The reason is the heavy traffic experienced on the street during daytime. Actually, anthropogenic heat generated by the traffic is also considered in the simulations for case 3 and case 4. But it is balanced by the predominating cooling impact of the roadside trees. In other words, the possible energy savings would be even larger in cases 3 and 4 if the heavy traffic was not considered.

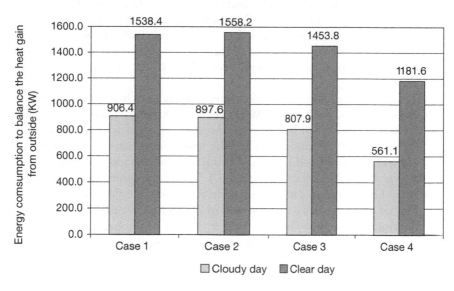

Figure 8.38 Comparison of cooling energy consumptions for the different cases.

Table 8.9 Potential energy saving caused by roadside trees.

	Case 1	Case 2	Case 3	Case 4
Cloudy day	0%	1%	11%	38%
Clear day	0%	–1%	5%	23%

Conclusion

1   The measurements conducted on a fair day indicate that housing estates with 22 per cent vegetation coverage had lower ambient air temperature compared to the one with only 7 per cent green coverage. The maximum average temperature difference between the two sites was 2.32 °C. The cooling benefit of greenery in-between buildings is obvious and it can contribute to a comfortable thermal environment and save cooling energy.

2   The average temperature difference between the business park (CBP) and its nearby traditional industrial area was 0.2 °C. The difference can be easily blurred by solar radiation during daytime in many ways: by placing strong radiation on plants and increasing ambient air temperature; by solar radiation being reflected from surrounding buildings with high-reflectivity façade materials; by solar radiation being absorbed by building structures during the daytime and releasing heat back to the

surroundings at night in the form of long-wave radiation. The thermal performance of the current landscape in the business park has its limitation. With more trees, it is expected to achieve a cooler environment. The evidence is that data obtained from the vacant area with many trees were consistently lower, from 1.4 to 1.8 °C, than the other locations.

3   The thermal condition in the industrial area is closely related to the density of roadside trees. With very extensive roadside trees, Tuas Avenue 2 performed best among the three selected streets. Tuas South Street 3 with very young trees had the worst performance. The long-term average temperature difference was 0.6 °C while the maximum difference in one particular day can be around 1 °C. On the other hand, anthropogenic heat can play a negative role in regulating the thermal conditions. Higher temperatures were observed in both Tuas Avenues 2 and 8 during daytime when the traffic was heavy there. The cooling energy savings caused by the roadside trees, up to 38 per cent, is very encouraging.

Case study IV
Intensive rooftop gardens

As one of the potential measures to mitigate the urban heat island effect, the utilization of plants for roof and sky-rise gardens has gained popularity around the world. In Singapore, the continuous effort to make the country a 'garden city' has already resulted in lush greenery throughout the island. It is necessary to demonstrate the thermal benefits of rooftop gardens in the tropics with tangible data. This is the primary reason why measurements and calculations were carried out locally.

Intensive green roofs, usually called 'rooftop gardens', are more commonly found in Singapore's local building developments. Two types of intensive rooftop gardens were involved in the case study. One is rooftop gardens on multi-storey carparks in the Punggol area and the other is a rooftop garden on a commercial building in the International Business Park.

Rooftop gardens in the Punggol area

Introduction

Two rooftops were involved in the measurement in the Punggol area, including a multi-storey carpark rooftop complete with garden landscaping, identified as C2 in the measurement, and another similar rooftop without any garden landscaping as yet, identified as C16 (see Figure 8.39). At C2, there are three types of profiles, which were defined as raised, sunken and exposed (see Figure 8.40). Both the multi-storey carparks consist of four storeys, with a rooftop that is accessible to the residents of the surrounding housing blocks. The experiment was carried out from 1 to 16 April 2001, a total of ten days. The month of April was chosen to represent the drier and hotter season of Singapore.

Discussion and observations

The locations with and without vegetation were first compared on a relatively clear day. The vegetation planted in the sunken bed was denser than that planted in the raised one during the measuring period. Figure 8.41 shows the

C2   C16

Figure 8.39 The two rooftops with and without landscaping.

Field measurement on rooftop garden

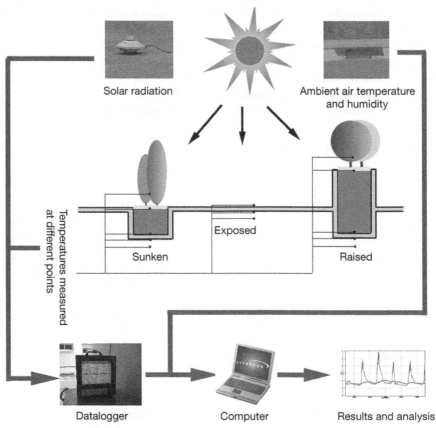

Figure 8.40 The measured profiles and the data collection process.

surface temperatures measured in the raised bed, the sunken bed and exposed concrete together against time. It is obvious that the locations shaded by plants had lower surface temperatures than the exposed one throughout the day. The maximal difference, around 10 °C, was observed at around 1330 hr. The surface temperature difference between the sunken bed and the raised one can also be observed due to the difference of density of plants. It is about 5 °C maximally at around 1330 hr.

The difference between the two sites with and without plants was of great interest. The long-term comparison (see Figure 8.42) indicates that the average temperature difference between the two sites, at 0.3 °C, is not very significant. However, a large variation in temperatures can be observed at site C16. In general, it has very high temperatures during the daytime and very low ones at night. Plants at site C2, however, controlled the deviation greatly. Therefore the temperature differences derived during daytime and night-time are not so obvious compared with those observed at C16. In order to have a clear look at the difference between C2 and C16, a clear day was chosen (see Figure 8.43). Over 6 °C temperature difference between C2 and C16 can be observed at around 1300 hr. It cannot be denied that the presence of plants has helped to bring down the temperature of the air, thus providing a cooler environment in the locality at C2. At night-time, the temperatures measured at C16 could be slightly lower than those measured at C2. With plants in the area, the process of evapo-transpiration occurs. This is reflected in Figure 8.43, where C16, having little or no plants around, exhibited lower RH values. During the afternoon when the temperature was at the maximum, the RH reduced significantly as the air in C16 could now hold more moisture.

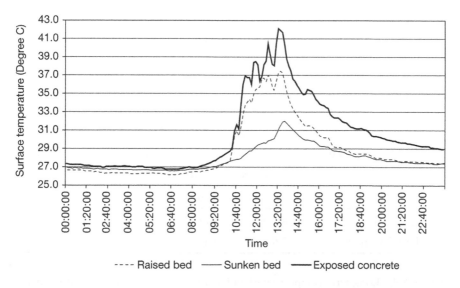

Figure 8.41 Comparison of surface temperatures measured under different profiles.

At C2, however, the RH did not decrease as much, with the rise in air temperature. Although the presence of plants brought about cooler temperatures, the side effect of high RH should be considered. The high RH will offset the better thermal comfort brought about by the lower ambient air temperature. Furthermore, the high RH may also have an impact on the building fabrics and materials as it may encourage biological growth and accelerate the degradation process. Therefore, it is essential that the rooftop gardens should be properly ventilated and that moisture can be removed in a timely fashion.

It is obvious that the presence of a rooftop garden can improve the harsh thermal environment of an exposed roof. It can directly contribute to the reduction of the surface temperature and indirectly contribute to the decrease of the ambient air temperature. On the other hand, the high relative humidity

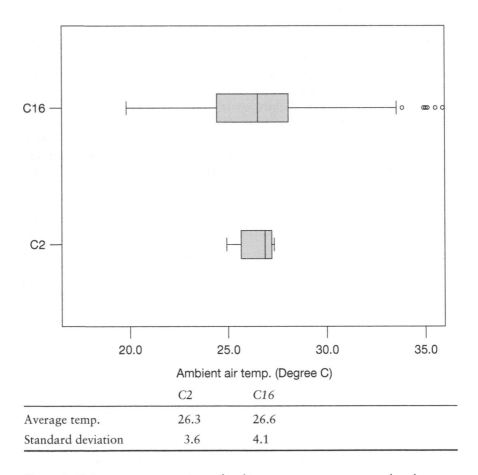

	C2	C16
Average temp.	26.3	26.6
Standard deviation	3.6	4.1

Figure 8.42 Long-term comparison of ambient temperatures measured at the two sites over the period from 1 to 12 April 2001.

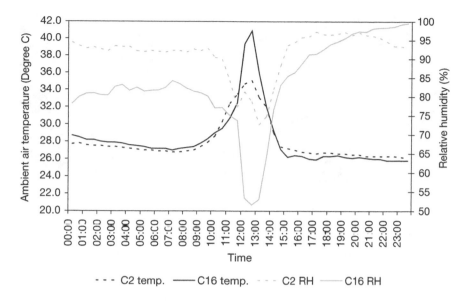

Figure 8.43 Comparison of ambient air temperature and relative humidity obtained at C2 and C16 on a clear day.

should be paid attention to and proper design should be applied to prevent moisture from accumulating on the rooftop.

Rooftop gardens in the International Business Park

Introduction

Another field measurement was carried out on an intensive rooftop above a low-rise commercial building (see Figure 8.44), which covers the planting of grass, shrubs and trees as well as pavement for access by visitors, in the International Business Park (IBP). The field measurement was not influenced by shadows or reflected solar radiation since there are no high-rise buildings around. Figure 8.45 shows the measuring points which were chosen on both planted and pavement areas and lasted from 25 October to 10 November, a total of 17 days. During the field measurement, the surface temperatures were measured in the interior and exterior environments respectively. At every measuring point, thermo-couple sensors were placed in close contact with the surface to capture surface temperatures. Ambient air temperature and relative humidity were also measured in the interior and exterior environments. For the outdoor environment, the ambient air temperatures and humidity were measured at different heights above vegetation and hard surface areas respectively. Besides the above items, globe temperatures, air

Figure 8.44 The rooftop garden on a low-rise building in the IBP.

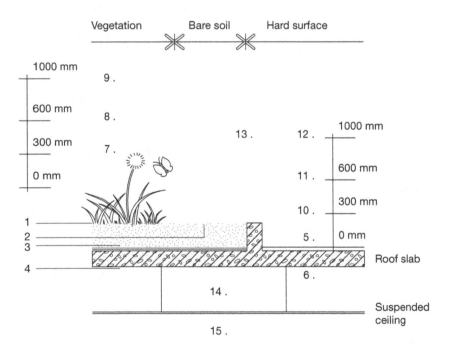

Figure 8.45 Measuring points set up on the rooftop garden.

velocity and solar radiation were measured on the rooftop as well. All the parameters were measured and recorded at 5-minute intervals. Based on the field measurement, an energy simulation was carried out subsequently in order to compare the possible energy savings brought by plants.

Discussion and observations

The direct effects of a planted roof are their thermal benefits in reducing the surface temperatures of roofs and the heat transfer into the rooms underneath. This will directly contribute to improving indoor thermal environment and the thermal performance of buildings. In the field measurement, the surface temperatures were measured under different species of vegetation with different Leaf Area Index (LAI) values. The aim was to compare their dissimilar abilities on reducing temperatures.

Figure 8.46 presents the comparison of surface temperatures measured with the different plants, bare soil and pavement area. Without plants, the maximum temperature of the hard surface could reach around 57 °C when solar radiation was at around 1400 W/m² around early afternoon. The daily variation of surface temperature on the pavement was over 30 °C. For bare soil, the surface temperature measured during the daytime was not as high as that of the hard surface. The maximum surface temperature of bare soil was around 42 °C and the fluctuation of the surface temperature throughout a day was around 20 °C. This could be due to evaporation of moisture in the wet soil that led to the reduction of surface temperatures during the daytime. With the presence of vegetation, the surface temperature was further reduced. The figure shows that the shading effect, or surface temperature reduction, of plants is highly dependent on the LAI since higher temperatures were usually found under sparse foliage, such as turfing, while lower temperatures were detected under dense ones, such as shrubs. The highest temperature measured under all kinds of plants was no more than 36 °C. Under the densest shrubs, the measured temperature was quite evenly distributed throughout the measurement. The maximum daily variation of surface temperature was no more than 3 °C and the maximum surface temperature was only recorded at 26.5 °C, which is much lower than those measured on the hard surface and the bare soil. The reduction of surface temperature was mainly caused by solar protection of plants. From the thermal benefit point of view, it is beneficial to have vegetation planted on the rooftop garden with a larger LAI like dense trees and shrubs. However, trees and shrubs could increase the roof structural load and maintenance. The selection of vegetation on the rooftop therefore requires a balance between these environmental, structural and maintainance considerations.

The direct thermal effect of plants was further evaluated by calculating the heat flux through both the exposed and the planted area on the roof (see Figure 8.47). The heat flux was calculated under conditions of bare roof,

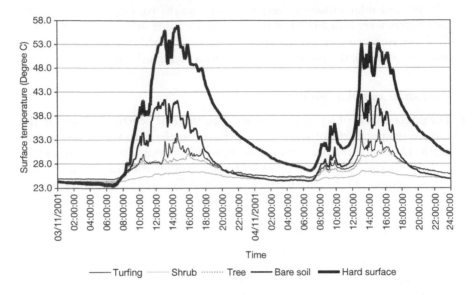

Figure 8.46 Comparison of surface temperatures measured with different kinds of plants, only soil and without plants on 3 and 4 November 2001.

bare soil (without any plants), turf, tree and shrub respectively. Compared with the planted areas, considerably higher heat flux was observed on the exposed roof throughout the whole day. The maximum heat flux, 19.76 W/m², was found at around 1400 hr. For the roof with bare soil or plants, the heat gain was subject to the different times of the day and the heat gain difference was observed to be obvious during the daytime. Plants played an important role in reducing thermal heat gain through their sun-shading effects during the period. Without shading, more heat gain was observed on the roof with only bare soil. At night, however, similar profiles were observed on both the areas with bare soil and plants. The results indicate that the heat gain during night-time was reduced mainly by the insulation effect of the soil layer. It is worth mentioning that total heat gain over the day was remarkably reduced to 297.2 KJ (around 40 per cent of the heat gain through the bare roof) with the introduction of a bare soil layer. Due to the presence of plants, the total heat gain over a day was further reduced. The minimum heat gain, around 164.3 KJ (around 22 per cent of the heat gain through the bare roof), was observed on the area with the dense shrubs.

Indirect effects of planted roofs refer to their potential thermal impact on the surrounding environment. They will contribute to creating a better outdoor thermal environment and will mitigate the UHI effect. Ambient air temperatures, humidity, globe temperature, mean radiation temperature (MRT) and reflected radiation measured above the hard surface and vegetation were therefore compared. The ambient air temperatures were measured

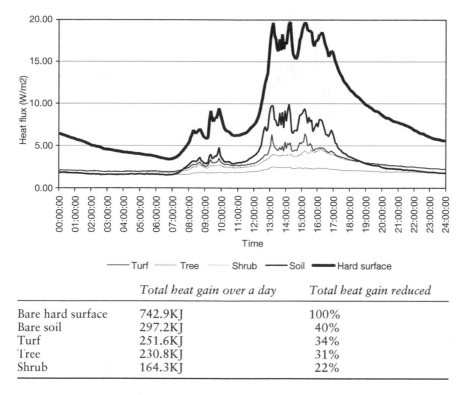

	Total heat gain over a day	Total heat gain reduced
Bare hard surface	742.9KJ	100%
Bare soil	297.2KJ	40%
Turf	251.6KJ	34%
Tree	230.8KJ	31%
Shrub	164.3KJ	22%

Figure 8.47 Comparison of heat flux transferred through different roof surfaces on 4 November 2001.

at different heights, 300 mm, 600 mm and 1 m, above the hard surface and the vegetation area (see Figure 8.48). In general, the ambient air temperatures measured near to the plants (300 mm) were about 4.2 °C lower than those measured near to the pavement area at around 1800 hr. But the difference was not there at the height of 1 metre above both the planted area and the pavement. For ambient air temperatures measured above both hard surface and vegetation, higher air temperatures were observed at lower heights during the daytime in general. This result indicates that both hard surface and vegetation, which were exposed to the strong incident solar radiation, had relatively high surface temperatures, which subsequently influenced the ambient air temperature by distance. With better air circulation at the higher positions, the ambient air temperatures measured at 1 metre above both hard surface and vegetation were obviously lower than that measured at lower levels during the daytime. After sunset, around 1830 hr, there was a significant reduction in ambient air temperatures above the vegetation. This provides evidence that vegetation can cool down easily and simultaneously can reduce the ambient air temperature at night. It is worth mentioning that

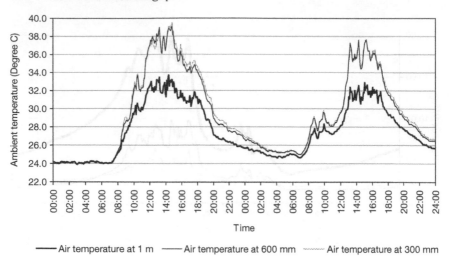

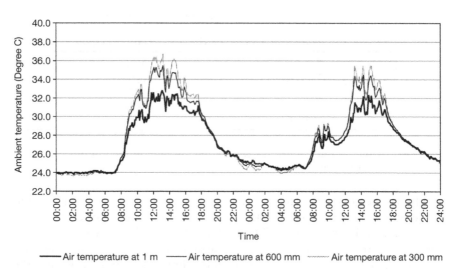

Figure 8.48 Ambient air temperatures measured at different heights above the hard
surface (top) and the planted area (bottom) on 3 and 4 November
2001.

the ambient air temperatures measured above the hard surface slowly
decreased after sunset. The ambient air temperature measured at 300 mm
height was always higher than those measured at higher levels even at night.
The absorbed heat was released back to the surrounding environment and
warmed the ambient air near to the hard surface throughout the day. This is
actually the root of the urban heat island effect in the built environment.

The comparison of humidity measured at 1-metre height above hard
surface and vegetation is shown in Figure 8.49. It seems that there was no

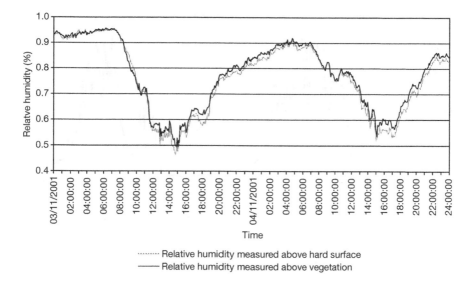

·········· Relative humidity measured above hard surface
————— Relative humidity measured above vegetation

Figure 8.49 Comparison of relative humidity measured with and without plants at 1m heights on 3 and 4 November 2001.

big difference between the measured humidity. This could be due to the close proximity of the measuring points. On the other hand, the good air circulation at the higher location can easily take the moisture away. This was reflected in the relatively low relative humidity observed above both hard surface and planted areas during the daytime.

Based on the measured ambient air temperatures, globe temperature and air velocity, MRT above the hard surface and vegetation were calculated respectively. The comparison of MRTs is illustrated in Figure 8.50. This shows an undisputed phenomenon – there were clear differences between MRTs calculated above hard surface and vegetation from sunset to sunrise of next day. The maximum difference of the MRT was 4.5 °C just after sunset (between 1800 and 1900 hr). Without direct sunshine, the radiative load mainly depends on the amount of long-wave radiation emitted from the surrounding surfaces. Since it was heated by solar radiation during the daytime, the hard surface had higher surface temperature and therefore emitted greater long-wave radiation to the surrounding environment at night. Green plants, on the other hand, could absorb part of the incoming solar radiation and protect the surface beneath from high surface temperature during the daytime. The long-wave radiation emitted from the vegetation, therefore, is much less than that emitted from the hard surface.

Figure 8.51 shows the comparison of measured values of the irradiated and reflected solar radiation from both hard surface and vegetation on 6 and 7 November. The measurement was conducted with solar meters hanging

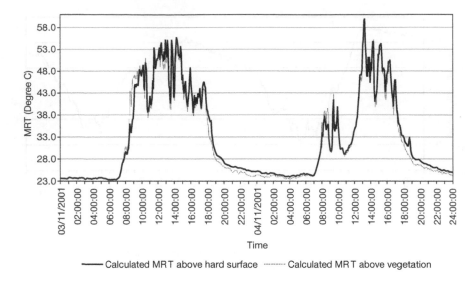

Figure 8.50 Comparison of MRTs calculated with and without plants at 1 m heights on 3 and 4 November 2001.

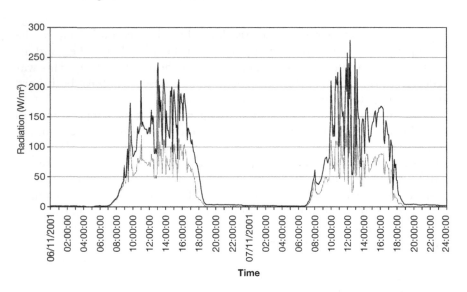

Figure 8.51 Comparison of irradiated and reflected radiation with and without plants at 300 mm heights on 6 and 7 November 2001.

upside down at 300 mm height above both hard surface and vegetation areas. Values measured above vegetation were absolutely lower than those measured above hard surfaces during the daytime. The maximum variation of 109 W/m^2 was found at noon when the incoming solar radiation was the strongest. This coincides with the view that the green plants irradiated and reflected less solar heat than the hard surface.

A comparison between rooftops without vegetation, rooftops covered by 100 per cent turfing, 100 per cent shrubs and 100 per cent trees respectively for a five-storey commercial building was carried out by the use of the PowerDOE energy simulation program. Two base cases (roofs with and without insulation) were taken into considerations in this comparison. The results are presented in Figure 8.52. It can be concluded that an insulation layer has a great impact on the heat transfer through a roof. All the figures show a large difference between the roofs with and without insulation. With insulation, the overall energy reduction is 29 MWH (19.5%). There is a significant reduction in the space cooling load and the peak space load of the typical flat roof compared to the exposed roof. For the five-storey building, the space cooling load reduced by 150.81 MWH (76.5%) and the peak space load reduced by 70.14 KWH (76.3%).

The comparison between the un-insulated roofs with and without rooftop gardens revealed that plants on an un-insulated roof had significantly reduced the heat gain into the building. The annual energy consumption was reduced by 19 MWH (9.5%) (rooftop covered by 100% turfing) to 29 MWH (19.5%) (rooftop covered by 100% shrubs) for the five-storey commercial building. The space cooling loads reduced by 92.94 MWH (47.1%) (rooftop covered by 100% turfing) to 155.85 MWH (79.0%) (rooftop covered by 100% shrubs), and the peak space load reduced by 43.1 KWH (46.9%) (rooftop covered by 100% turfing) to 72.48 KWH (78.9%) (rooftop covered by 100% shrubs). The reductions in the annual energy consumption with the installation of the rooftop gardens on an un-insulated rooftop imply that the overall running cost of the building will decrease.

The comparison between the well insulated roofs with and without rooftop gardens revealed that plants can still reduce the heat gain but the reductions were not as significant. The results derived from the simulation indicate that a roof equipped with both rooftop garden and insulation cannot double its energy saving. Instead, replacing the insulation layer with plants should be considered. On one hand, an un-insulated roof with rooftop garden can have a similar or even better energy consumption pattern compared with an insulated roof without plants. On the other hand, unlike an insulation layer which just rejects heat gain into individual buildings but burdens the surrounding environment, a green roof can actually benefit not only the building underneath but also the environment.

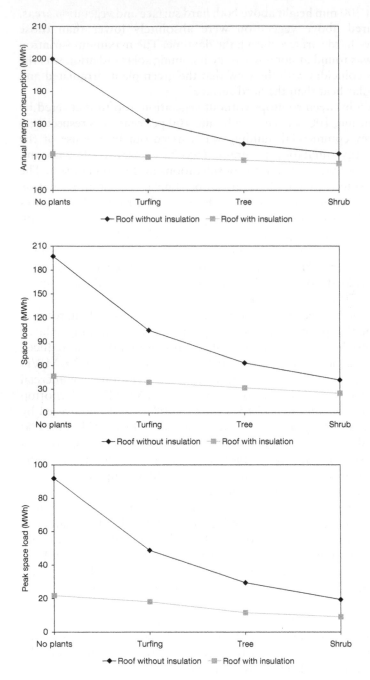

Figure 8.52 Comparisons of annual energy consumption, space load component (total building load) and peak space load component (total building load) for different types of roofs for a five-storey commercial building.

Conclusion

In this case study, the direct and indirect thermal impacts of intensive rooftop gardens in a tropical climate were investigated. Some quantitative data are summarized as follows:

- With the shading of plants, surface temperatures measured under different kinds of vegetation were much lower than that measured on the hard surface. The maximum temperature decrease caused by plants was around 30 °C. The temperature measured under vegetation varied according to the density (LAI) of plants. Normally, lower temperatures were measured under thick foliage while higher temperatures were obtained under sparse vegetation or only soil.

- The heat transfer through the bare roof was greater than that through planted roofs and a roof with only soil. Compared with the bare roof, much less heat gain was observed on the planted roof. Actually, the thermal benefits caused by the planted roof were the combined effects of both soil layer and planted vegetation. Wet soil can provide an additional insulation effect to the roof for the whole day and vegetation mainly provides sun protection during the daytime.

- The cooling effect of plants was confirmed by ambient air temperatures measured at different heights. The maximum temperature difference of 4.2 °C was detected with and without plants on site. But the cooling effect was limited by distance.

- The humidity measured on the rooftop with plants was normally higher than that without plants. On the other hand, good ventilation on the rooftop can dissipate the moisture. Therefore, proper design with the layout of the roofs and surrounding buildings should be carefully considered.

- Less long-wave radiation emitted from the planted roof was confirmed through comparisons of globe temperature and MRTs measured on site. Maximum differences of the globe temperature and the MRT were 4.05 °C and 4.5 °C respectively just after sunset. This gives evidence that planted roofs could effectively mitigate the urban heat island effect in urban environments.

- That green plants irradiated and reflected less solar heat was confirmed by the measurement of reflected solar radiation on site. The maximum variation of 109 W/m^2 was found at noon.

- The installation of a rooftop garden on a five-storey commercial building had resulted in a saving of 0.6–19.5 per cent in the annual energy consumption, 17.0–79.0 per cent in the space cooling load and 17.0–78.9 per cent in the peak space load.

Case study V
Extensive rooftop gardens

Unlike intensive rooftop gardens, extensive green roofs are not designed for public access but are developed mainly for aesthetic and ecological benefits with relatively low initial cost, lightweight and thin growing media and minimal maintenance. In Singapore, more and more extensive rooftop gardens have been installed on the existing buildings in order to achieve aesthetic and environmental benefits.

Rooftop gardens in the Punggol area

Introduction

To explore the thermal impacts of four different extensive rooftop greenery systems, a *before* and *after* measurement was carried out on a multi-storey carpark in the Punggol area (see Figure 8.53). The designated rooftop, with two decks at different levels, was empty when the *before* measurement was carried out. The first phase of measurement was carried out from 19 May to 9 June 2003 over a period of 22 days. The *after* measurement was done at the same locations but around eight months later when the whole rooftop (including the upper and the lower decks) was totally planted with four different green systems (see Figure 8.54), from 14 February to 3 March 2004. The rooftop was divided into four equivalent areas represented by G1, G2, G3 and G4 to accommodate the four different systems in the measurement. The objective was to explore four green roof systems under tropical conditions, and to evaluate if significant differences in thermal performance existed between the systems.

Discussion and observations

The surface temperature is a major indicator which can determine the thermal performance of the measured objects. The surface temperatures were measured on the exposed rooftop and on the areas subsequently covered by different types of extensive rooftop gardens. Extra measuring points were selected on the surface of the substrate. The comparisons (3 to 4 June 2003

Figure 8.53 Designated measuring points on the bare roof (before) and the green roof (after).

vs. 22 to 23 February 2004) of four individual areas (G1, G2, G3 and G4) are presented from Figure 8.55 to Figure 8.58. Compared with surface temperatures measured on the exposed rooftop during the *before* session, the surface temperatures of the rooftop measured during the *after* session were much lower. The fluctuations were also minimal (within 4 °C). This is understandable because of the thermal protection of the holistic system, waterproof layer, drainage layer, rooting materials, vegetation and so on covered above these locations.

However, compared with surface temperatures measured on the exposed rooftop during the *before* session, the surface temperatures measured on the substrates but under different vegetations were higher mostly during the daytime. The maximum surface temperature could be up to around 60 °C. This is amazing since the surface temperatures were measured under vegetation rather than on the exposed surface. Possible reasons for such high surface temperatures are:

1   The substrates are mostly dark in colour which means they will absorb more solar heat and easily incur higher surface temperatures.

Residential buildings

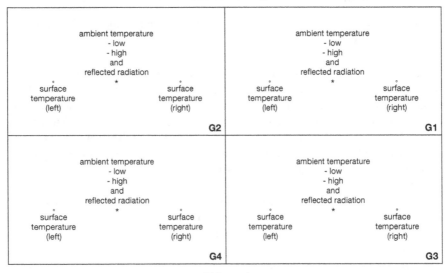

| ambient temperature
- low
- high
and
reflected radiation
*
surface   surface
temperature   temperature
(left)   (right)

G2 | ambient temperature
- low
- high
and
reflected radiation
*
surface   surface
temperature   temperature
(left)   (right)

G1 |
| ambient temperature
- low
- high
and
reflected radiation
*
surface   surface
temperature   temperature
(left)   (right)

G4 | ambient temperature
- low
- high
and
reflected radiation
*
surface   surface
temperature   temperature
(left)   (right)

G3 |

Main road

Figure 8.54 The four extensive systems installed on the rooftop.

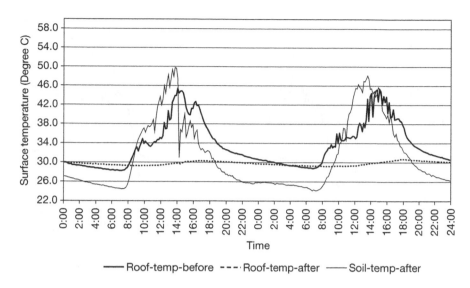

Figure 8.55 Comparison of surface temperatures measured on G1.

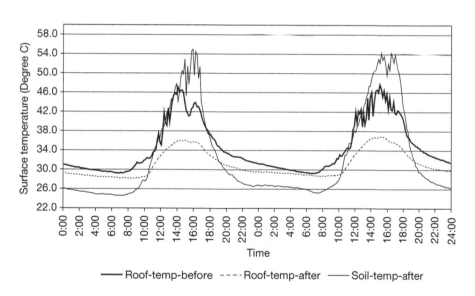

Figure 8.56 Comparison of surface temperatures measured on G2.

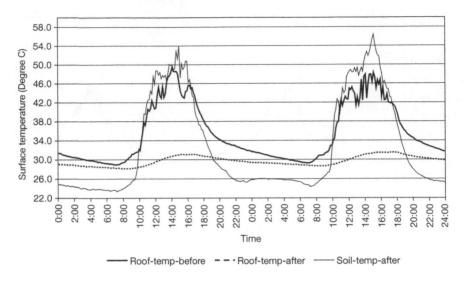

Figure 8.57 Comparison of surface temperatures measured on G3.

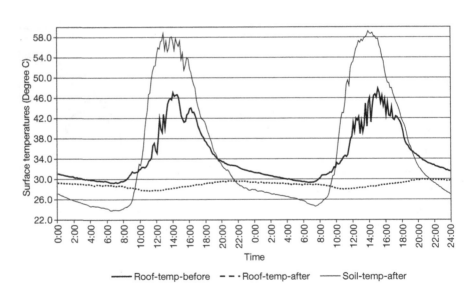

Figure 8.58 Comparison of surface temperatures measured on G4.

2   The thermal capacities of the substrates of the extensive roof garden systems are rather small compared with the intensive system. With a thin layer and lightweight substrate, the heat can easily build up during the daytime and dissipate at night. This is the reason why the fluctuation of surface temperatures of the substrate surface is so large.

3   The vegetation growing on the rooftop gardens is not very extensive. Therefore, the thermal protection is limited.

4   A prolonged drought period was experienced before the *after* measurement, therefore the substrate was very dry and the evaporative cooling effect was marginal.

Besides the above-surface temperatures, another two surface temperatures measured on the exposed surface were monitored during the *after* session as reference points (see Figure 8.59). A cross-comparison of substrate surface temperatures among all the four individual areas and two reference points on the exposed surface was done. During the daytime, it is amazing that all surface temperatures measured under vegetation were higher or similar to surface temperatures measured on reference point 1 (exposed surface) except for the point measured on G1, which was occasionally lower. The reasons have been explained earlier. It seems that the thermal protection from vegetation is marginal. However, some temperature differences can still be observed under different types of plants. It is obvious that the density of vegetation governs the surface temperatures measured underneath. G2, G3 and G4 are all ground cover with tiny leaves. The surface temperatures measured under them (especially G4) were higher. G1 has relatively extensive greenery. The temperatures measured there were relatively lower.

To further explore the thermal performance of four extensive rooftop gardens with higher substrate moisture, data derived from the period 29 February and 1 March 2004 were compared with those obtained from the *before* session. The thermal performance of rooftop greenery systems on 29 February and 1 March was obviously better than that on 22 and 23 February 2004. Maximally, over 18 °C of surface temperature decrease was observed in G3 around 1400 hr. Extremely high surface temperature of the substrate did not occur at this time. This could be explained by the combined effects of lower ambient air temperature and higher substrate moisture. Similar observations were obtained from the comparisons of surface temperatures of substrates and exposed surface. All locations within the systems experienced lower surface temperatures. Compared with the extreme scenario (measured on 22 to 23 February 2004), the extensive rooftop greenery performed well on some days when the substrate was not very dry. The maximum difference of surface temperature between exposed surface and substrate surface was around 20 °C which is observed in G3. In terms of foliage density, G3 is not the best case. Therefore, the lowest surface temperature observed in G3 indicates that it is not caused by the shading effect of vegetation alone but by the whole system including the substrate.

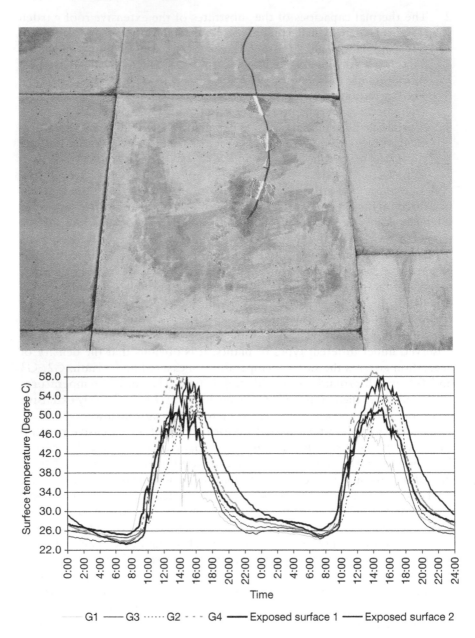

Figure 8.59 Comparison of substrate surface temperatures with exposed surface temperatures.

The heat flux through the concrete slab under different systems was calculated according to the above and soffit surface temperature difference (see Figure 8.60 to Figure 8.63). The positive values indicate that the heat flux is from above to below the slab while the negative ones move the opposite way. Generally, the heat flux is minimum during the night-time and maximum during the daytime when there was no roof greenery. This is understandable since the temperature difference between the roof surface and soffit is high during the daytime while it is marginal at night. After implementing the rooftop greenery, inverse patterns occurred on the rooftop. Due to higher surface temperature on the roof slab, more heat transfer occurs through the slab at night. This is because the heat is not easily dissipated due to the protection of the extensive systems. On the other hand, less heat was transferred from above to below the slab during the daytime with the protection of those extensive systems. Some negative values which indicate the opposite transfer of heat could be observed during the daytime. All these are owing to the excellent thermal protection of the extensive systems which can create surface temperatures of the slab without much fluctuation.

To further compare the heat passed through the slab, the calculated heat gain/loss per square metre over a clear day (22 February 2004) is presented in Table 8.10. It is clear that overall heat gain through the slab is greatly reduced due to the installation of extensive systems. G3 performs very well in terms of preventing heat gain. Over 60 per cent of heat gain could be reduced. In terms of thermal protection obtained by vegetation, G3 is not the best case. Probably, the combined effect of the whole system, especially the polyurethane layer used as a water drainage and water reservoir layer for

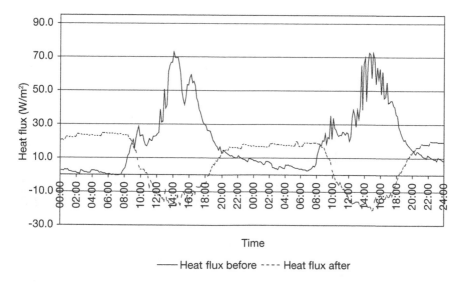

Figure 8.60 Comparison of heat flux through the concrete slab on G1.

this particular green roof system, causes less heat to be transferred through the system.

For every individual rooftop garden, ambient air temperatures were measured at different heights during the *before* and the *after* sessions. The comparisons of ambient temperatures measured on 3 and 4 June 2003 and

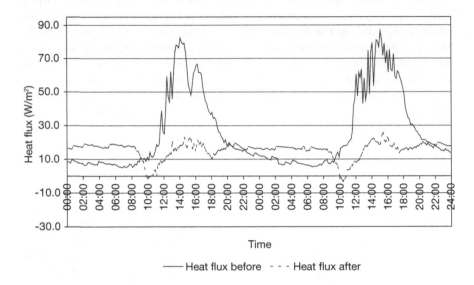

Figure 8.61 Comparison of heat flux through the concrete slab on G2.

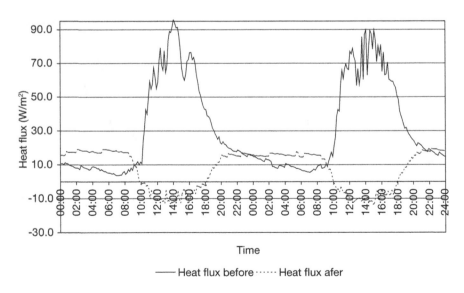

Figure 8.62 Comparison of heat flux through the concrete slab on G3.

22 and 23 February 2004 are presented from Figure 8.64 to Figure 8.67. Basically, the closer to the surface, the higher the ambient air temperatures were experienced during the daytime. This occurred during both *before* and *after* sessions but it was more intense during the *after* session. Furthermore, peak ambient air temperatures measured above vegetated areas were all higher than those measured above exposed hard surface although weather data showed relatively close ambient air temperatures for both *before* and *after* sessions during the daytime. The high surface temperatures of the substrate may have a predominant impact on ambient air temperatures measured at different heights. During the daytime, the substrates work like a 'heat source'. The weak evapo-transpiration effect of vegetation has been totally blurred.

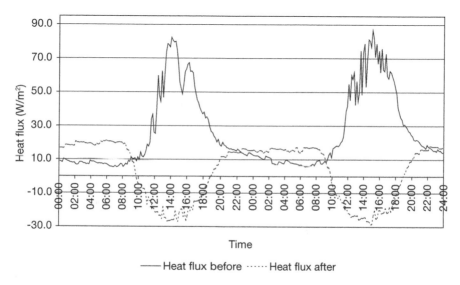

Figure 8.63 Comparison of heat flux through the concrete slab on G4.

Table 8.10 Comparison of total heat gain/loss over a clear day (22 February 2004) on the rooftop before and after.

	Total heat gain/m² over a day (KJ)	Total heat loss/m² over a day (KJ)
G1 before	1681.3	0.9
G1 after	1072.0	301.8
G3 before	2638.9	0
G3 after	864.6	213.1
G2 before	2079.1	0
G2 after	1335.7	1.2
G4 before	2117.0	0
G4 after	864.5	561.7

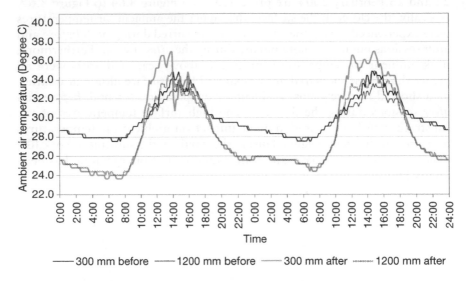

Figure 8.64 Comparison of *before–after* ambient air temperatures measured in G1 (3 and 4 June 2003 vs. 22 and 23 February 2004).

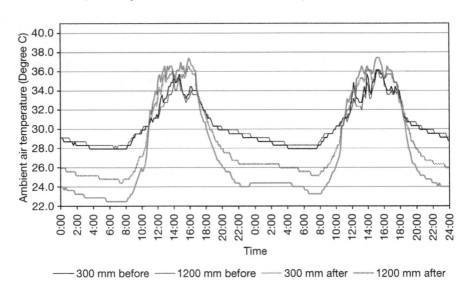

Figure 8.65 Comparison of *before–after* ambient air temperatures measured in G2 (3 and 4 June 2003 vs. 22nd and 23rd February 2004).

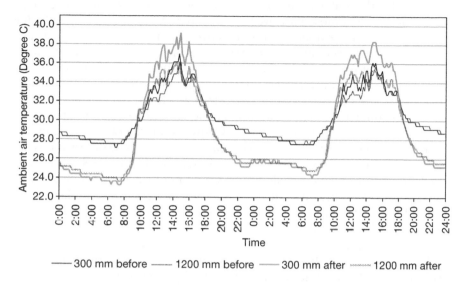

Figure 8.66 Comparison of *before–after* ambient air temperatures measured in G3 (3 and 4 June 2003 vs. 22 and 23 February 2004).

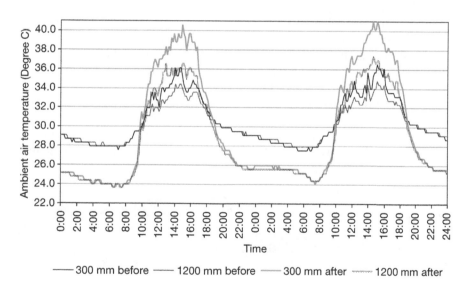

Figure 8.67 Comparison of *before–after* ambient air temperatures measured in G4 (3 and 4 June 2003 vs. 22 and 23 February 2004).

At night, the ambient air temperatures measured above vegetated areas are all lower than those measured above exposed roofs. This is mainly due to the change of weather conditions during the two sessions. It was also observed that the cooling effect of plants was minimal through temperature profiles except for G2 where the temperature difference during the daytime was not much for different heights and it became obvious at night. Here the inverse situation occurred in that the closer to the vegetated areas, the lower the ambient air temperatures. The lowest ambient air temperatures were observed, which was around 1 °C lower than the others. This means that vegetation was working as a 'cooling source'. All these observations are due to the relatively extensive greenery in G2 (see Figure 8.68). However, such an inverse situation could not be found in other areas where ambient air temperatures measured at different heights were quite similar all the time. The worst scenario was found in G4. At heights of both 300 mm and 1200 mm, the maximum temperature difference between *before* and *after* sessions during the daytime were all observed at G4. They were 6.1 and 3.8 °C respectively. The worst situation may be caused by the dark-colour substrate, the high surface temperatures and sparse vegetation. The performance of the other two extensive systems (G1 and G3) were between G2 and G4. There is no clear evidence showing the cooling effects of plants during both daytime and night-time.

In general, the cooling effect of all four extensive systems was minimal in terms of reducing ambient air temperature. The closer to the vegetated area, the higher the ambient air temperature observed during the daytime. This indicates that the cooling effect of vegetation, if there was any, had been blurred by the substrates with high surface temperature. At night, ambient air temperatures measured at different heights were more or less the same. Only vegetation in G2 showed an obvious cooling effect.

The nature of the plants used on the green roofs could also have an impact on the thermal performance of the green roofs. The majority of the plants

Figure 8.68 Different conditions of G2 (left) and G4 (right) during the measuring period.

used were those using the Crassulacean Acid Metabolic (CAM) mode of photosynthesis. Accordingly, it is expected that many plants will not transpire during the daytime, and therefore transpiration cooling by plants will be minimal during the daytime. The distribution of CAM plants in the various systems was as follows:

- G1: about 10 per cent
- G2, G3, G4: almost 100 per cent.

CAM plants are supposed to fully or partially close their stomata in order to maintain high rates of photosynthesis during the daytime. Therefore, the cooling effect should be obvious for G2, G3 and G4 during the night-time rather than the daytime. The cooling effect could be observed in G2 and G3 but not in G4 at night. There was no cooling effect observed in G1 during the daytime as well. The reason is probably that G1 and G4 had less extensive greenery and the effect of vegetation was marginal.

The reflected global radiation from the rooftop was measured at all four plots. Basically, reflected radiation measured during the *after* session was lower than that measured during the *before* session. This could be explained by the fact that the original smooth concrete surface was replaced by extensive rooftop gardens. Less direct and diffuse reflected radiation could be measured. However, we could not conclude that long-wave radiation emitted from the rooftop had been decreased. On the contrary, long-wave radiation emitted from vegetated areas may be increased since high surface temperatures of substrates were experienced during the daytime.

G4 (see Figure 8.69) had the best scenario since more than 50 per cent of peak reflected radiation had been decreased after the installation of the extensive system. This is probably due to the darker colour of the substrate. Radiation was absorbed rather than reflected. The performances of the rest of the systems were similar and no more than 30 per cent of peak reflected radiation had been reduced.

The comparisons of reflected global radiation measured on 3 and 4 June 2003 as well as 29 February and 1 March 2004 indicate that the reflected global radiation measured during the *after* session was still lower than that measured during the *before* session for all four systems. Due to lower surface temperatures of substrates, the reflected global radiant measured was further decreased on 29 February and 1 March 2004 compared with that measured on 22 and 23 February 2004. G4 was still the best plot where on average 39.2 per cent of original global radiation was reflected.

To visually compare the performances of the four systems, a group of infrared pictures was taken on 1 April 2004 and 3 November 2004 respectively (see Figure 8.70 to Figure 8.71 (see also colour plates). On 1 April 2004, the difference between G1 and G3 could easily be observed (see Figure 8.70; see also colour plates). Basically, the surface temperature of G3 was lower than G1 when the plot was fully covered by ground cover. However,

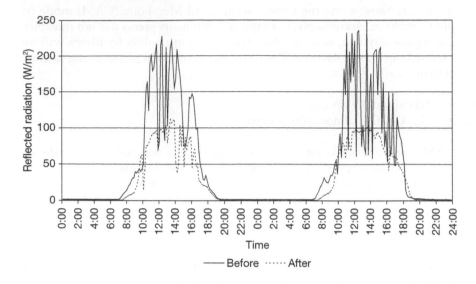

Figure 8.69 Comparison of reflected global radiation measured at G4 (3 and 4 June 2003 vs. 22 and 23 February 2004).

the surface temperature could be very high (even higher than the exposed concrete surface) in G3 where the substrate was exposed without greenery. The maximum temperature difference between the well-planted area and exposed substrate could be up to 20 °C. G1 was covered by clusters of shrub and the substrate was not well covered at some locations where higher surface temperature was experienced. On 3 November 2004, G1 was well covered by the garden system compared with the situation observed on 1 April 2004 (see Figure 8.71, see also colour plates). Therefore, the thermal performance of G1 in terms of surface temperature is better than that of G3. There is not much difference in G3 between the system observed on 1 April 2004 and 3 November 2004.

The differences between G2 and G4 can be observed in Figure 8.72 and Figure 8.73 (see also colour plates). On 1 April 2004, G2 was better than G4 in terms of greenery coverage. The plot of G2 was well shaded by vegetation and its surface temperature was relatively lower. On the other hand, G4 was mostly exposed and its surface temperatures were high. On 3 November 2004, G4 was well covered by plants and exhibited a good performance although it was still not as good as G2.

The visual comparisons of four roof systems indicate that the coverage of greenery plays an important role in achieving good thermal performance. Lower surface temperature was observed with more extensive greenery. On the other hand, systems with substrates exposed to strong sunshine may incur high surface temperature and cause worse thermal conditions during the

Figure 8.70 Comparison of G1 and G3 (1 April 2004). (See also colour plates)

Figure 8.71 Comparison of G1 and G3 (3 November 2004). (See also colour plates)

Figure 8.72 Comparison of G2 and G4 (1 April 2004). (See also colour plates)

Figure 8.73 Comparison of G2 and G4 (3 November 2004). (See also colour plates)

daytime. Sometimes the surface temperature of exposed substrate was even higher than the exposed concrete surface. Since low maintenance strategy has been implemented for all four extensive systems, the growing condition of vegetation should be a major concern for choosing proper extensive systems in the future.

Rooftop gardens on a metal roof

Introduction

The use of metal roofing is common in local industrial buildings. However, it was found in the previous macro measurement that metal roofing can induce high surface temperatures and therefore cause daytime UHI effects near the industrial area. It is of great interest to explore the performance of greenery on metal roofs. Due to their limited loading capacity, metal roofs can probably only accommodate the extensive rooftop system. A metal roof covered with vegetation was selected. Three types of plants were measured on the metal roof. The measurement was carried out from 3 September to 21 October 2005.

Discussion and observations

The data obtained on the green metal roof has been analysed over a long period (see Figure 8.74). The impact of density (LAI values) on the fluctuation of surface temperatures can be clearly noted from the variations observed from dense plants, sparse plants and weeds. The mean surface temperatures ranged from 27.1 °C to 30.4 °C and the standard deviations ranged from 1.4 to 3.5. The benefits of reducing the surface temperature with greenery can be observed from mean surface temperature differences between the hard metal surface and that below the plants. They are 4.7 °C, 1.9 °C and 1.4 °C with the presence of dense plants, sparse plants and weeds respectively. The benefits of greenery were also reflected in controlling the surface temperature fluctuation. Without plants, the metal surface can reach a temperature of up to 60–70 °C during the daytime (observed from the high whisker of the hard surface) and lower than 20 °C at night (observed from the low whisker of the hard surface). Even the 50 per cent observations are distributed between 23 °C to 39 °C. With plants, the maximum surface temperature observed below the dense plants was around 32 °C and the minimum one was around 24 °C. The reduction of standard deviation can also be observed with the presence of plants.

To further uncover the pattern during the daytime, a long-term analysis was carried out based on the data recorded during the daytime (from 0700 to1900 hr, see Figure 8.75). A significant increase in terms of the mean surface temperatures can be observed on the hard surface. It was up to 7.8 °C while only 0.3 to 1.6 °C increases were observed under the plants. The metal is very sensitive to the variation of incoming solar radiation.

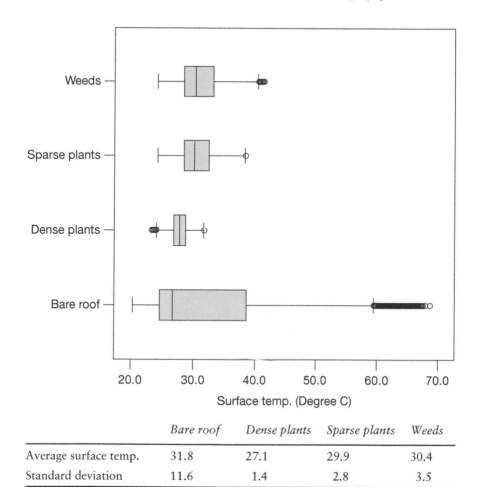

	Bare roof	*Dense plants*	*Sparse plants*	*Weeds*
Average surface temp.	31.8	27.1	29.9	30.4
Standard deviation	11.6	1.4	2.8	3.5

Figure 8.74 Long-term analysis of the surface temperatures measured above the green metal roof.

Therefore, its surface temperature can be very high during the daytime although it drops dramatically at night. Green roofing adds a buffer to the high sensitivity roof in this manner. It can effectively reduce the fluctuation of the surface temperatures. The contribution is appreciable during the daytime when the UHI effect can be easily triggered by the hard surfaces in a built environment. In Singapore, most low-rise industrial buildings employ metal roofing which burdens the cooling energy consumption by transferring a large proportion of heat gain from the roofs. The extensive rooftop garden can definitely benefit industrial buildings by reducing the cooling energy use.

Subsequently the observation focused on a clear day when the performance of plants can be found easily (see Figure 8.76). It can be observed that the

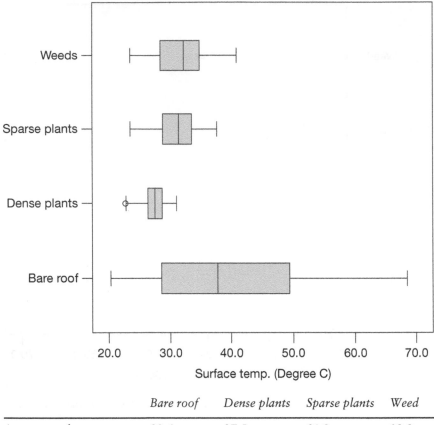

	Bare roof	*Dense plants*	*Sparse plants*	*Weed*
Average surface temp.	39.6	27.5	31.2	32.0
Standard deviation	11.9	1.6	3.1	4.0

Figure 8.75 Long-term analysis of the surface temperatures measured above the green metal roof (excluding night-time) from 0700 to 1900 hr.

surface temperatures of the exposed metal roof are very sensitive to solar radiation. They more or less followed the profile of solar radiation during the daytime. The peak value was around 60 °C which was observed at 1230 hr when the solar radiation also reached its peak value. The maximum difference between the surface temperatures of the exposed metal roof and below the dense plants was 35.1 °C during the daytime. At night, the inverse situation occurred and the surface temperatures of the exposed metal roof dropped dramatically. They were lower than all the temperatures measured below the rooftop greenery. The maximum surface temperature difference between the exposed metal roof and the dense plants was 4.74 °C at night. A comparison of surface temperatures measured on the metal roof was also done on a cloudy day (see Figure 8.77). Differences can still be observed during the daytime. The surface temperatures of the exposed metal roof can be very low with the decrease of solar radiation during the daytime. But they are still higher than those measured under the dense plants. A maximum

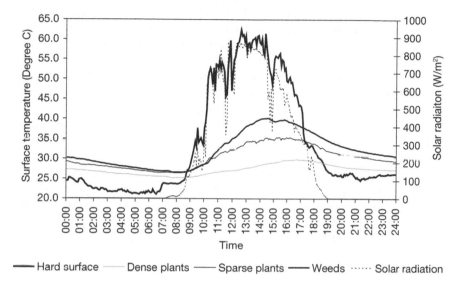

Figure 8.76 Comparison of surface temperatures measured on the green metal roof on a clear day (16 September 2005).

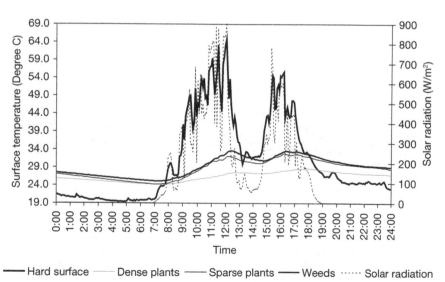

Figure 8.77 Comparison of surface temperatures measured on the green metal roof on a cloudy day (3 October 2005).

temperature difference of nearly 5 °C can still be detected between the dense plants and the other plants and the hard surface during the daytime.

The correlation analysis of solar radiation and the surface temperatures measured at different locations is shown in Figure 8.78. It can be observed that the surface temperature of the exposed metal surface is the most sensitive variable to the solar radiation incident on the roof. On the other hand, the surface temperature measured under the dense plants is the least sensitive one due to their outstanding sun-protection effect. Of course, the sequence of the gradients follows the density of plants.

Infrared pictures were also taken for the purpose of visually comparing the variation of surface temperature over a planted metal roof during the daytime (see Figure 8.79; see also colour plates). Generally, both surface temperatures of the exposed metal roof and the extensive system increased until 1500 hr. However, with the protection of the plants, the fluctuation of the surface temperature of the green roof was very little compared with that of the exposed roof. On the other hand, the metal hard surface can cool down at a faster pace compared with the green roof after 1700 hr. All the observations support the view that metal roofs are very sensitive to the fluctuation of solar heat during the daytime and induce very high surface temperatures. Such sensitivity can be modified by green roofs in terms of reducing the very high surface temperature during the daytime. This totally accords with the observations obtained from the previous field measurement.

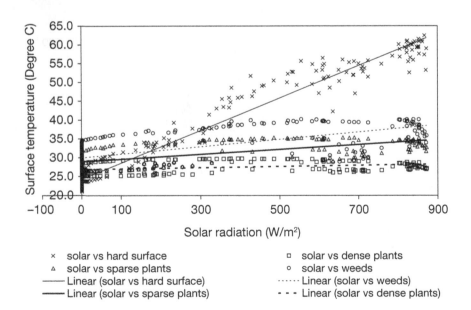

Figure 8.78 Correlation analysis of solar radiation and the surface temperatures on a clear day (16 September 2005).

0825 hr

1010 hr

1200 hr

1500 hr

1700 hr

1900 hr

Figure 8.79 Infrared pictures of the metal roof with greenery on 5 December 2005.
(See also colour plates)

The cross-comparison of the planted roof and the exposed roof further reveals the difference between the bare metal roof and the green roof during the daytime around 1330 hr (see Figure 8.80; see also colour plates). All the hard surfaces, such as the metal roof, pavement and asphalt, showed red to pinkish colour, which means high surface temperature in the spectrum. They are all heat sources which are sensitive to solar radiation during the daytime. Heat is massively absorbed in this manner and is reflected or emitted to the environment later. The green roof, however, showed relatively low temperature as indicated by the greenish colour when the solar radiation was at its peak during the day. This indicates the outstanding sun-protection effect of the green roof. Further, less heat will be released back into the environment eventually.

Conclusion

Extensive green roofs tend to experience a lower surface temperature compared to an exposed roof surface, especially in areas well covered by vegetation. The maximum temperature differences of 18 °C and over 30 °C were observed on the concrete roof and the metal roof respectively. However, the effect of the extensive systems may not be as remarkable as the intensive ones, since:

Figure 8.80 Cross-comparison of the planted roof and the exposed roof.

- a thin layer of substrates which represent low thermal capacity result in high surface temperatures caused by the reradiated heat. On the other hand, substrates with low thermal capacity can be easily cooled down at night, and therefore less heat is trapped and reradiated;
- some substrates are dark in colour with the tendency to absorb more heat during the daytime;
- coverage of greenery in the system is of great importance. It is believed that better thermal performance of the extensive system can be achieved when the substrate is fully covered with plants;
- the plants used in green roofs are mostly low-lying ground cover types of plants, and therefore did not yield results as significant as the measurements made on intensive-type rooftop gardens.

Other positive impacts of the extensive rooftop garden were:

- heat flux through the roof structure was greatly reduced due to the installation of extensive systems. Maximally, over 60 per cent of heat gain was stopped by the system;
- the risk of glare for surrounding buildings was reduced due to the installation of extensive systems.

Overall, these extensive rooftop garden systems, due to their light weight and easy maintenance, are convenient to be built on existing buildings, especially metal roofs. They may bring many thermal as well as environmental benefits. However, their indirect thermal effects on the immediate surroundings are not comparable to those of intensive systems.

Case study VI
Vertical landscaping

Compared with green roofs, vertically placed greenery can cover more exposed hard surfaces in an urban environment, especially where high-rise buildings are predominant. In addition, thermal, visual and acoustical benefits and the improvement of air quality can be obtained by this means. Vegetation planted vertically is not a new concept. With rich vegetation species and abundant rainfall in the tropical climate, vertically grown plants can be observed both in nature and in the built environment (see Figure 8.81). However, strategically introducing plants into building façades is still a challenge due to the lack of R & D in this area. In order to explore the impacts of vertical planting in the region, some preliminary measurements have been carried out.

Figure 8.81 Vertically grown plants can be observed in nature and in the built environment.

Sun protection by vertically placed plants in the vicinity of housing developments

Introduction

One of the important features of plants is their ability to dissipate incoming solar radiation. As solar energy is needed for photosynthesis and growth, most plants can adjust the angle of their leaves to the direction of the sun's maximum radiation (Deering, 1953). Therefore, plants can provide the optimum sun-shading effect for vertical façades. To explore such sun-shading effects, the impacts of vegetation planted in different places (pergola, overpass, fencing, etc.) were measured in a public-housing neighbourhood. The field measurements were carried out on some clear days in May 2000. During these periods, the domestic solar azimuth was northerly and the mean daily solar radiation was 424.4 mWh/cm^2.

Discussion and observations

The climber is one of the favourite species to be introduced into the vertical sides of buildings. The solar radiation was measured vertically in front of and behind climbing plants on two pergolas at different orientations at 60-minute intervals during daytime hours from 0830 hr to 1730 hr. The results are shown in Figure 8.82 and Figure 8.83. For Pergola A, peak values of solar radiation without plants covering SE and NE directions were detected at 1030 hr, around 350 W/m^2 and 625 W/m^2 respectively. Compared with the original incoming solar radiation, the solar radiation measured behind climbing plants remained low and even. These values at the SE and the NE directions were no more than 55 W/m^2. For Pergola B, measuring results were quite similar to Pergola A although the species of plants were different. This indicates that climbing plants can efficiently block the incoming solar radiation. After finishing measurements, leaves within certain areas were collected from two pergolas to calculate the Leaf Area Index. The LAI of the climbing plants on Pergola A was 1.9 while that on Pergola B was 3.6. With more foliage, trees could have higher LAIs than those of the climbing plants. However, data derived from measurements indicated that sparse plants, such as that planted on the Pergola A, could already decrease the incoming solar radiation on vertical surfaces efficiently. This provides evidence that climbing plants can be an effective 'sun-shading device' on vertical sides of buildings.

Shrubs are also suitable for introduction into buildings for decreasing the solar radiation on vertical sides. Clusters of shrubs planted in front of a substation were chosen to measure the sun-shading effect of shrubs. The gap between the shrubs and walls was 0.5 m wide. The measurements were carried out at the north, the east and the west orientations respectively at 60-minute intervals during daytime hours from 0800 hr to 1800 hr and the results are shown in Figure 8.84. With maximum values of incoming solar

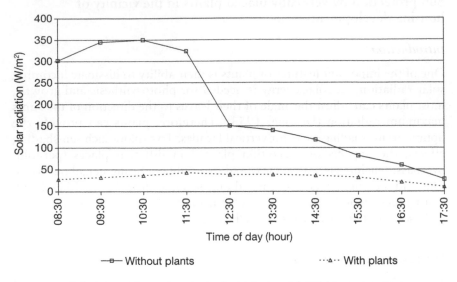

(a) Solar radiation reading taken with plants facing the SE orientation for Pergola A.

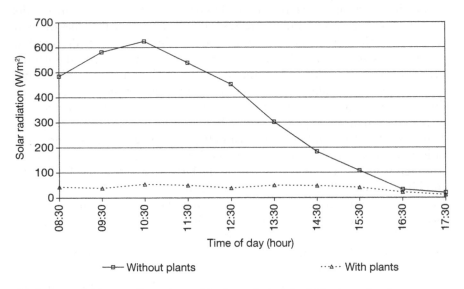

(b) Solar radiation reading taken with plants facing the NE orientation for Pergola A.

Figure 8.82 The sun-shading effect of climbing plants on Pergola A.

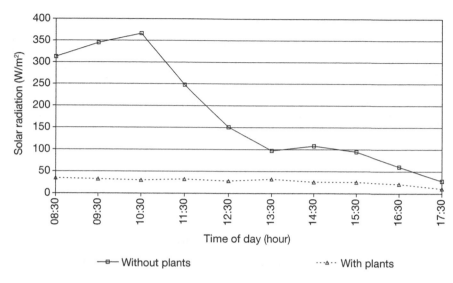

(a) Solar radiation reading taken with plants facing the SE orientation for Pergola B.

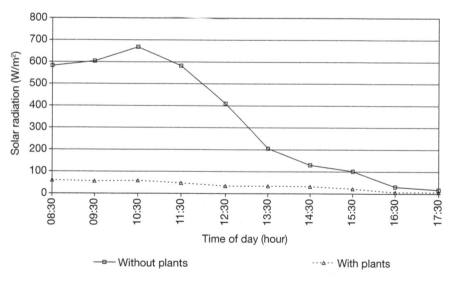

(b) Solar radiation reading taken with plants facing the NE orientation for Pergola B.

Figure 8.83 The sun-shading effect of climbing plants on Pergola B.

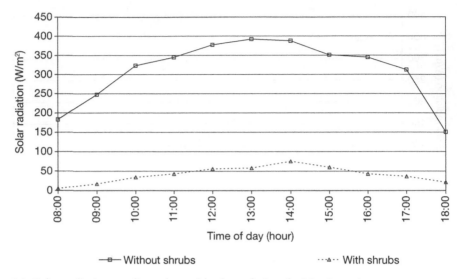

(a) Solar radiation reading taken with plants facing the N orientation.

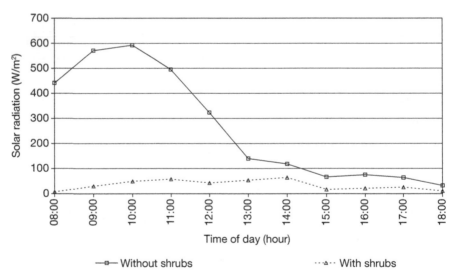

(b) Solar radiation reading taken with plants facing the E orientation.

Figure 8.84 The sun-shading effects of shrubs on walls of the substation.

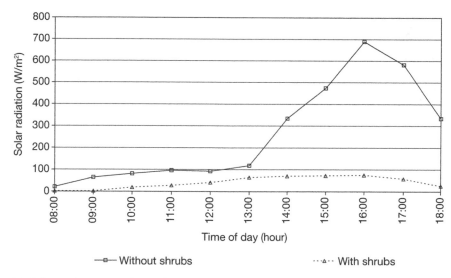

(c) Solar radiation reading taken with plants facing the W orientation.

Figure 8.84 Continued

radiation striking the walls without shrubs, 593 W/m² and 690 W/m² were detected in the early morning (1000 hr) and late afternoon (1600 hr) at the east and the west orientations respectively. For the north orientation, the highest incoming solar radiation was detected at 1300 hr, around 392 W/m². On the other hand, the values of solar radiation measured behind the shrubs at these three orientations were relatively low and even, no more than 75 W/m². The highest solar radiation behind shrubs was detected around 1400 hr at the north orientation. Similarly, the solar radiation behind shrubs reached its peak around noon at the east and the west orientations. The reason could be that the shrubs could not cast sufficient shade on the wall due to the high solar altitude around noon. To get better sun-shading effects, the gap between plants and walls should be sufficiently narrow for the foliage to cast shade efficiently on walls.

Bushes planted on overpasses are common in Singapore. The green plants decorate the harsh overpasses. On the other hand, employing bushes in the air indicates that plants can be introduced onto the hard surfaces of tall buildings. An overpass with bushes was chosen to measure its sun-shading effect at 60-minute intervals during daytime hours from 0830 hr to 1830 hr. Similar to previous measurements, the solar radiation can be efficiently decreased by plants (see Figure 8.85). The solar radiation measured behind plants remains low and stable even during the early morning and late afternoon.

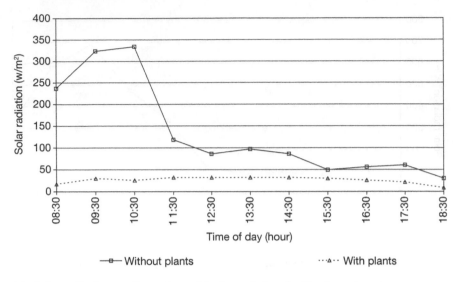

(a) Solar radiation reading taken with plants facing the E orientation.

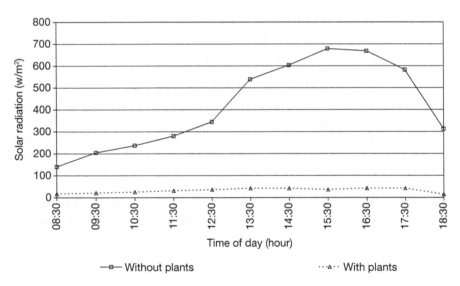

(b) Solar radiation reading taken with plants facing the W orientation.

Figure 8.85 The sun-shading effects of bushes on the overpass.

Trees planted near to façades

Introduction

Genuine vertical planting, especially that introduced onto the upper parts of buildings, is very rare in Singapore. Although trees planted near to buildings are not the direct vertical planting method, they can provide valuable reference data on the subject. Some measurements were carried out as follows:

- trees planted at the different orientations of residential blocks;
- trees planted at the western orientation of industrial buildings.

Discussion and observations

1. Trees planted at the different orientations of residential blocks

To capture the surface temperature decrease caused by trees on the west side, a residential block with western corridors was selected (see Figure 8.86). The measuring object was a dark green parapet. The surface temperature was measured on the third floor. The measuring point a could be shaded by trees

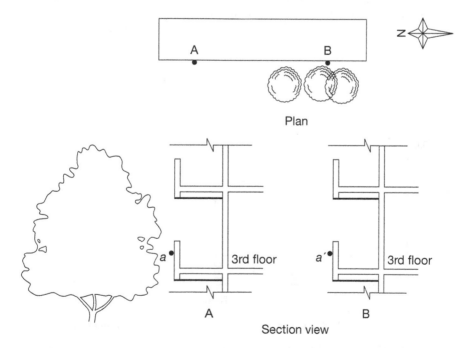

Figure 8.86 A 15-storey slab block with west corridors.

while *a´* was exposed to the direct sunlight. From Figure 8.87 we see that the surface temperature increase was significant from 1300 hr to 1700 hr without trees. On the other hand, the surface temperature measured under the shade of trees was much lower. The maximum difference of surface temperatures, around 11 °C, was detected at 1700 hr. It was observed that the outstanding temperature difference was normally detected on the darker surface. To some extent, the colour of the vertical side decides its absorption property of solar radiation, which influences the surface temperature. The darker the colour, the higher the absorption. On the other hand, the absorption of solar radiation by the dark wall behind the thick plant covering is negligible. So the surface temperature difference between darker-coloured material with or without plant shading would be greater than the temperature difference produced by lighter-coloured material. In other words, the thermal effect of plants seems more 'effective' when introduced to a dark-colour surface.

The white east parapets of a 15-storey condominium were measured to capture the thermal effects of trees on the east side. Trees were very close to the parapet. The air temperature, surface temperature, humidity and wind speed were measured on the 2nd floor, the 5th floor, the 10th floor and the 15th floor. The 5th floor was at the tree-top level. The plan of the condominium and the position of each measuring point are shown in Figure 8.88. With trees, the surface temperatures measured on the outside and inside surfaces of the parapet were quite similar (see Figure 8.89). The internal side of the parapet was not exposed to sunlight and the surface temperature measured there was not directly influenced by solar radiation. Similar surface

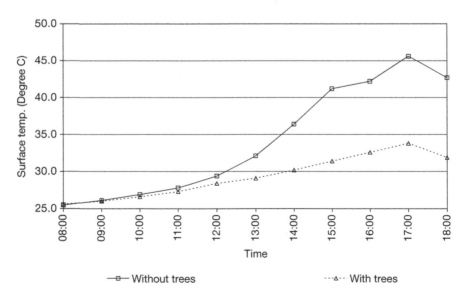

Figure 8.87 Comparison of surface temperatures measured on west parapet, with and without trees (clear day, 25 September 1999).

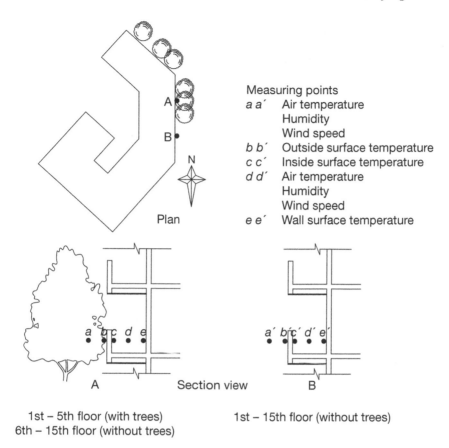

Measuring points

a a´	Air temperature
	Humidity
	Wind speed
b b´	Outside surface temperature
c c´	Inside surface temperature
d d´	Air temperature
	Humidity
	Wind speed
e e´	Wall surface temperature

Plan

N

Section view

A

B

1st – 5th floor (with trees)
6th – 15th floor (without trees)

1st – 15th floor (without trees)

Figure 8.88 Plan of the condominium and the position of each measuring point.

temperatures measured on both sides indicated that there was little heat flow caused by strong solar radiation through the parapet due to the protection by trees. On the other hand, without the shading cast by trees, the surface temperatures of the external and the internal sides of the parapet were quite different. The difference caused high heat flow from outside to inside.

Figure 8.90 shows the comparison of the surface temperature, humidity and air temperature measured at *b, d* and *b´, d´* on the 5th floor. Both surface temperatures and air temperatures under the tree covering were lower than those without trees. The maximum surface temperature difference, around 5 °C, was detected at 1300 hr. With trees, the air temperature within the corridor was around 1.5 °C lower than that without trees. However, the humidity was higher under the shade of trees. These measurements indicate that trees can reduce the surface temperature and cool the surrounding air. On the other hand, they simultaneously increase the humidity. Minimizing

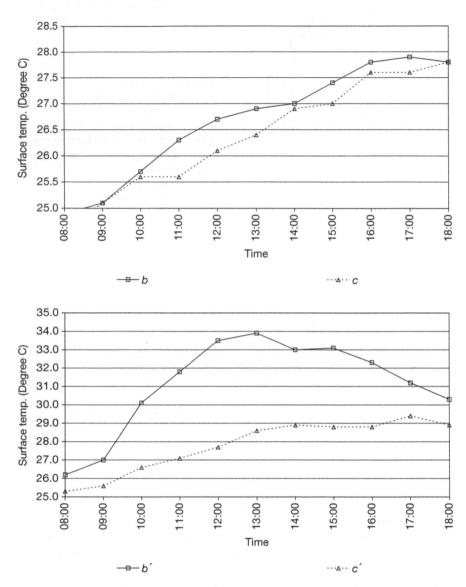

Figure 8.89 Relationships between surface temperatures measured on the external and the internal sides of the parapet (clear day, 13 January 2000).

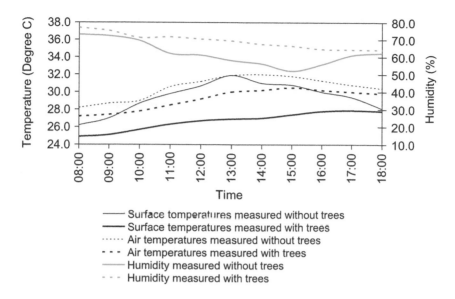

Figure 8.90 Comparison of the surface temperature, humidity and air temperature with and without trees (clear day, 13 January 2000).

wind blockage, therefore, should be one of the major considerations when we introduce plants into local buildings.

2. Trees planted at the western orientations of industrial buildings

A measurement was carried out in the industrial area in Singapore. Two factories with and without trees planted at western orientations were measured in a woodlands link. The façade colour for the two factories was dark blue.

A long-term comparison of the temperature variations with and without tress is shown in Figure 8.91. The outstanding shading effect of trees is reflected by a narrow span of temperature variation. It can be observed that the reduction occurs mainly at the max-whiskers which should be detected during the daytime when solar radiation is strong. The trees can effectively intercept the incident solar radiation and generate a lower surface temperature behind them on the façade. On the other hand, no significant difference can be observed between the two min-whiskers. This indicates that the impacts caused by trees at night are not obvious. The mean surface temperature recorded behind the trees was 28.7 °C while that recorded on the exposed façade was 30.1 °C.

In order to have a close look at the performance of trees on reducing the surface temperatures on the western orientation, two days were selected. Figure 8.92 shows a comparison made on a relatively clear day. The shading

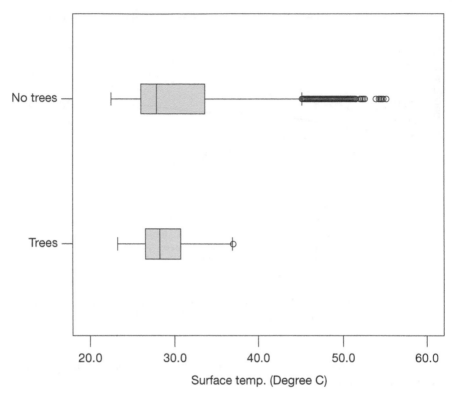

Figure 8.91 A long-term comparison of the surface temperature variations with
and without trees from 21 September to 7 December 2000.

effect caused by trees over the western orientation can be easily observed
during the daytime. Due to the orientation effect, there is a time lag between
the peak of solar radiation and the peak of external surface temperatures
measured on the western façades. The maximum temperature difference can
be up to 13.6 °C at around 1530 hr. Figure 8.93 shows a comparison made
on a relatively overcast day. Without much direct radiation, the shading effect
of trees is mainly reflected in reducing the diffused radiation on the spot.
A temperature difference of 7 °C can still be observed at around 1550 hr.
This finding highlights the impact of vertical shading on building façades not
only on clear days but also in overcast conditions. On the other hand, the
surface temperatures are inversely distributed at night compared with those
observed during daytime. The difference is constantly around 2 °C. Without
the blockage of the foliage, the heat can be easily dissipated by the exposed
façade to the surroundings. Compared with the stunning reduction of surface
temperature during daytime, the limitation that occurs at night can be
neglected.

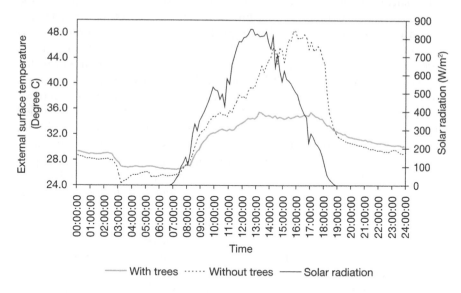

Figure 8.92 Comparison of solar radiation and the surface temperatures measured with and without shading from trees on 1 November 2005.

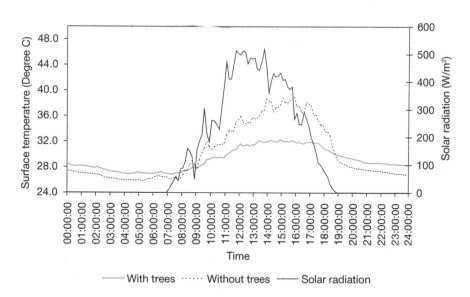

Figure 8.93 Comparison of solar radiation and the surface temperatures measured with and without shading from trees on 15 November 2005.

The surface temperatures were measured externally and internally at the second storey in two factory buildings. At this level, Factory 7 is partially shaded while Factory 11 is still fully protected by trees. The comparison is shown in Figure 8.94. It is obvious that the internal surface temperatures measured in the two buildings are more or less the same throughout the day (with a consistent deviation of less than 1 °C). The external temperatures, however, are quite different during the daytime. The maximum difference can be up to 9.2 °C. At night, the surface temperatures of the partially shaded surface are consistently lower (around 0.5 °C) than those of the well shaded one. The findings highlight the importance of shading caused by plants on vertical façades. Without shading, the maximum difference can be up to 13.6 °C. It drops to 9.2 °C when the façade is partially shaded. Well shaded by plants, the maximum internal–external difference is only about 4 °C. The figure can be up to 13.7 °C when the façade is not well shaded.

The possible reduction of heat flux through the well shaded façade with reference to that through the partially shaded one during the daytime on a clear day is shown in Figure 8.95. It is assumed that the buildings are naturally ventilated. The average heat flux through the well shaded wall is only 57 per cent of that through the partially shaded one. Further reduction can be imagined when the well shaded façade is compared with an exposed one. A comparison of Delta T (surface temperature difference) on 1 November 2005 is shown is Figure 8.96 (F11 is well shaded while F7 is partially shaded by trees). It is obvious that the two factories were in the process of heat loss at night and heat gain during the daytime. More heat

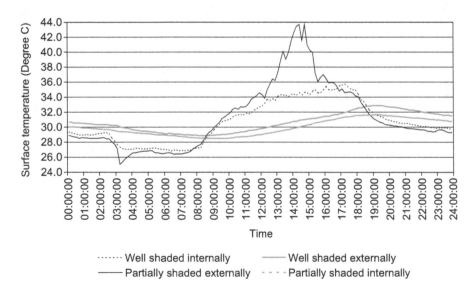

Figure 8.94 Comparison of the surface temperatures measured at second storey on 2 November 2005.

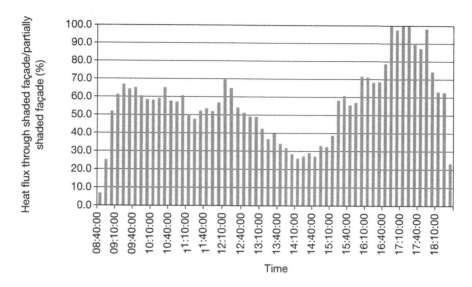

Figure 8.95 The reduction of heat flux through the well shaded façade with reference to that through the partially shaded one during the daytime.

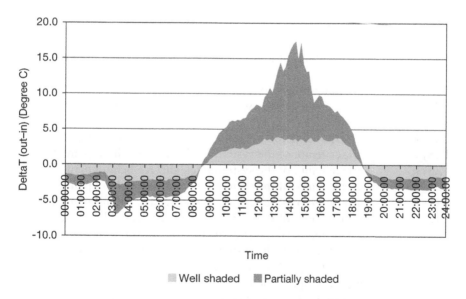

Figure 8.96 A comparison of Delta T (surface temperature difference) on 1 November 2005 (F11 is well shaded while F7 is partially shaded by trees).

can be dissipated at night in F7. But this cannot compensate the high heat gain during the daytime. It is important to note that most cooling energy is consumed during the daytime to meet the requirement of various activities.

Vertical greenery system

Introduction

A genuine vertical greenery system was installed on 5 June 2007 (see Figure 8.97). The dimension of the system is 2.41 m (width) by 4.25 m (height) and it covers roughly two stories on the western facing wall of a three-storey building. Behind the system, there are a naturally ventilated store room and an air-conditioned lecture room at the first and the second storey respectively. Four types of plants and two types of growing media were employed in the vertical system. According to these variables, surface temperatures were measured behind the system (at the back of growing media), behind different types of plants, on the exposed locations and indoors for long-term monitoring of the performance of the vertical greenery system. Altogether ten groups of measuring points were set up and every measuring group included four points (see Figure 8.98). The legends of every measuring point are summarized as follows:

1st:	1st storey
2nd:	2nd storey
SAWall:	surface temp. of the wall behind the vertical system with soil type A
NatSA:	corresponding indoor point of SAWall at 1st storey
SBWall:	surface temp. of the wall behind the vertical system with soil type B
NatSB:	corresponding indoor point of SBWall at 1st storey
P1Sys:	surface temp. of the system back surface behind plant 1
P2Sys:	surface temp. of the system back surface behind plant 2
P3Sys:	surface temp. of the system back surface behind plant 3
P4Sys:	surface temp. of the system back surface behind plant 4
P1:	surface temp. of the system surface protected by plant 1
P2:	surface temp. of the system surface protected by plant 2
P3:	surface temp. of the system surface protected by plant 3
P4:	surface temp. of the system surface protected by plant 4
Ex:	surface temp. of the bare wall
NatEx:	corresponding indoor point of Ex at 1st storey
AirconSA:	corresponding indoor point of SAWall at 2nd storey
AirconSB:	corresponding indoor point of SBWall at 2nd storey
AirconEx:	corresponding indoor point of Ex at 2nd storey

Figure 8.97 The vertical greenery system installed on the western wall of a local building.

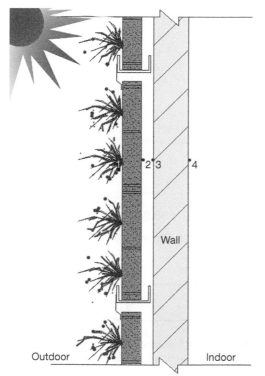

Figure 8.98
Section view of one group of measuring points across the system and the wall.

Discussion and observations

A comparison over a long period from 5 June to 28 August 2007 is presented in Figure 8.99. The highest temperature fluctuation can be observed on the bare wall, followed by that on the growing media behind the plants. It is very obvious that the variation of the surface temperatures of the wall behind the vertical greenery system (including both the plants and the growing media) is relatively small throughout the period. The ability to maintain the surface temperatures at a certain range is also reflected by the values of the standard deviation. Both surface temperatures measured behind the system and behind the plants have smaller standard deviations compared with that of surface temperatures observed on the bare wall. Overall, judging from the left and

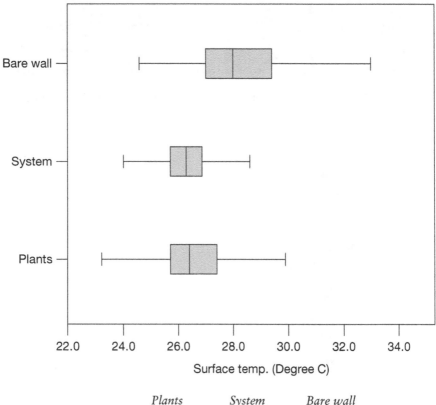

	Plants	*System*	*Bare wall*
Average temp.	26.6	26.3	28.3
Standard deviation	1.5	0.8	2.0

Figure 8.99 Comparison of surface temperatures measured over a period of time on the bare wall, behind the vertical greenery system and behind the plants.

the right whiskers, it can be observed that the vertical greenery system can induce lower surface temperature during both daytime and night-time. But the absolute difference between the temperatures obtained from the bare wall and those measured behind the plants and the system, around 2 °C, is not very much. This can be explained by the orientation of the wall. Since it is western facing, the bare wall is not warmed up by the sun until after noon when a big difference between the walls with and without the vertical greenery system can be observed.

In order to have a close look at the performance of the vertical system on a clear day, 27 June 2007 was selected to do a 24-hour analysis. Figure 8.100 shows a comparison of the wall surface temperatures measured on the bare area, behind the system and indoors at the first storey (naturally ventilated environment). The temperature differences can be clearly observed between the protected points and the bare ones. The maximum difference is around 7–8 °C which can be observed at around 1800 hr (late afternoon). The indoor–outdoor surface temperature difference behind the vertical greenery system is minimal, within 1 °C only. This indicates a minimal heat gain during daytime due to the protection of the vertical greenery system. However, the indoor–outdoor difference at the exposed area is obvious, at around 4 °C maximally. Meanwhile, diurnal fluctuation of indoor–outdoor surface temperature behind the system is also very small, at around 1 °C. Throughout the day, the vertical system can consistently maintain a fairly low surface temperature behind it. This indicates the good thermal protection provided by the system (note that it is the function of the system, not the

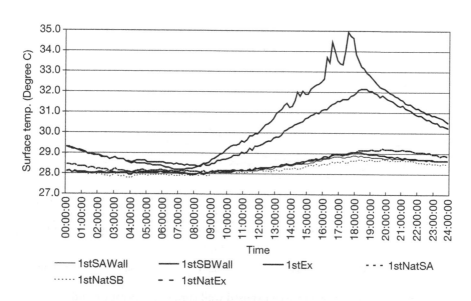

Figure 8.100 Comparison of wall surface temperatures measured at the first storey (27 June 2007).

plants alone). In terms of different types of growing media (soil types A and B), there is no significant influence on the surface temperature.

Figure 8.101 shows the comparison between exposed wall surface temperature and temperatures of the areas shaded by different types of plants. In general, the surface temperatures on the bare wall are mostly higher than those of the planted area. However, the surface temperatures measured behind certain plants can exceed the surface temperatures of the bare wall occasionally, such as plant 1 in this case. This can probably be explained by the dark colour of the growing media and the sparse arrangement of the plants. The irrigation system works during the early morning and early evening. At around 1700 to 1800 hr, the dark colour peat moss dried up easily because of inadequate solar protection from the sparse plants. Coupled with very strong solar radiation, a very high surface temperature can be experienced. During the night-time, the surface temperatures of the planted areas were consistently lower (up to 4 °C) than that of the bare surface. This is encouraging since it means that less heat will be released back to the environment at night when the hard surface is covered by the vertical greenery system.

Figure 8.102 shows the comparison of the wall surface temperatures measured on the bare area, behind the system and indoors at the second storey (air-conditioned environment). Again, surface temperature differences can be clearly observed between the planted points and the bare ones. The maximum difference is around 8 °C. The indoor–outdoor surface temperature difference behind the system is still small but it is not as small

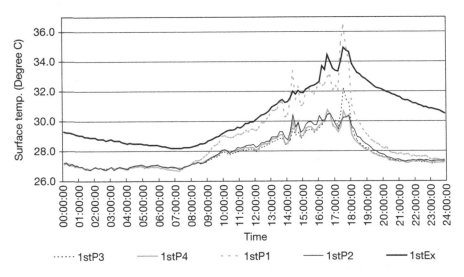

Figure 8.101 Comparison between exposed wall surface temperature and temperatures of the areas shaded by different types of plants at the first storey (27 June 2007).

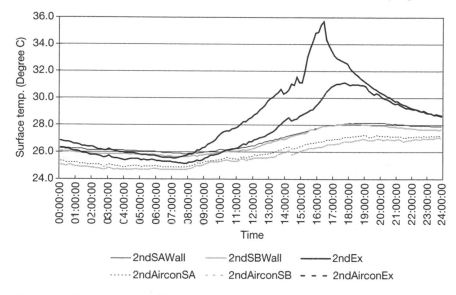

Figure 8.102 Comparison of wall surface temperature at the second storey (27 June 2007).

as that which appeared at the first storey due to the influence of air conditioning indoors. On the other hand, the indoor–outdoor surface temperature difference on the exposed area is obvious, at around 4 to 5 °C maximally even with air conditioning indoors. Such heat flux from outdoor to indoor can definitely consume more cooling energy. The diurnal fluctuation of indoor–outdoor surface temperature behind the system is still small, at around 1 °C.

Figure 8.103 shows the comparison between exposed wall surface temperature and temperatures of the areas shaded by different types of plants at the second storey. From 0000 hr to 1100 hr, the surface temperatures of the bare wall were lower than those measured behind the plants. The possible reason for this may be due to the influence from the air conditioning inside. For the rest of the time, surface temperatures behind the plants were lower than those obtained from the exposed wall although the difference is not so remarkable. Higher surface temperature can still be observed behind the sparse plant 1.

Conclusion

Vertical planting is not popular in Singapore at the moment. However, the preliminary findings derived from the field measurements are encouraging:

- Plants can effectively intercept incident radiation, up to around 80–90 per cent, near the vertical surface.

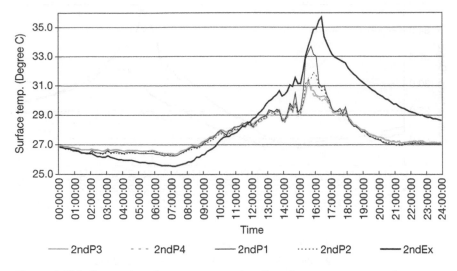

Figure 8.103 Comparison between exposed wall surface temperature and temperatures of the areas shaded by different types of plants at the second storey (27 June 2007).

- Trees planted near a building can efficiently reduce the surface temperatures of hard surfaces. The temperature reduction is very much dependent on the density of trees and the orientation of façades. Up to about 15 °C temperature reduction can be observed on eastern or western facing façades protected by dense trees.
- A vertical greenery system can provide excellent thermal protection for walls. The relatively low surface temperature of the vertical greenery system throughout the day can also contribute to the environment by releasing less long-wave radiation.
- The entire temperature reduction can be translated into cooling energy saving for buildings and environmental benefits for the built environment. More research is necessary in this area in the future.

Reference

Deering, R. B. (1953). *Technology of the Cooling Effect of Trees and Shrubs.* Washington, DC: Building Research Advisory Board.

Case study VII
Green experiments

From the previous field measurements carried out by the authors, it has been found that some parameters, such as the Leaf Area Index, are critical in terms of deciding the thermal performance of plants in the built environment. However, these critical parameters are not controllable in the field measurements. In order to explore the correlation between LAI values and the corresponding thermal performance, two control green experiments were set up.

Experiment on a rooftop

Introduction

A control experiment was built up on a rooftop. The basic setting of every experimental box included a 1 m by 1 m by 1 m steel frame, a 1 m by 1 m by 0.1 m concrete slab on top, insulation material (polyethylene and aluminum foil) and rolling wheels (see Figure 8.104). The insulation was applied at four vertical façades and the bottom side. One side of the concrete slab was exposed for further treatment in the experiment.

The experiment was carried out on the rooftop of the National University of Singapore, Department of Building. The boxes were placed at reasonable intervals without interfering with each other. There were two types of experimental boxes. One was intensive and the other had extensive settings. The major difference between the intensive and the extensive settings was the thickness of the soil layer. The intensive setting had 300 mm thick soil layer and a 4 mm insulation rubber. Bigger shrubs could be planted. On the other hand, the extensive setting had 100 mm soil layer, 1 mm filter layer, 50 mm drainage layer, 3 mm protection layer and 1.8 mm waterproofing system. The extensive setting was suitable only for the planting of lightweight turf rather than dense plants. Three types of plants were employed in the experiment (see Figure 8.105). A sparse red shrub with LAI of 1.5 and a dense green shrub with LAI of 5.5 were planted on the tops of the intensive boxes. The extensive one was planted with turf and its LAI value was approximately 3.3. For control purposes, a box with bare soil and a box with

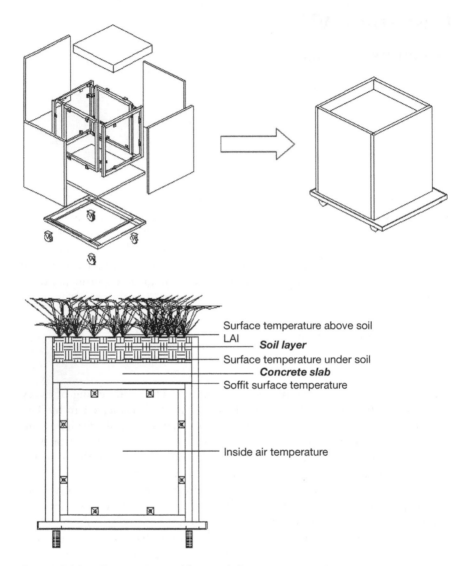

Figure 8.104 The experimental box and the measuring points.

Surface temperature above soil
LAI
Soil layer
Surface temperature under soil
Concrete slab
Soffit surface temperature

Inside air temperature

exposed concrete were also built up. Simultaneously, continuous weather data were monitored by a weather station installed on the rooftop.

There were two phases in the measurement. In the first phase of the experiment, the insides of the experimental boxes were not air-conditioned but sealed with insulation. In the second phase, the comparison was done when cool air was pumped into the boxes through an air conditioner to maintain a constant low internal temperature.

Figure 8.105 The three plants employed in the experimental boxes.

Discussion and observations

The experiment was carried out from 13 July to 2 September 2003, a total of 52 days. A typical day was chosen to highlight the temperature profile. The soil surface temperatures under plants and the bare concrete temperature are compared in Figure 8.106. As expected, the presence of plants had a huge impact on the soil surface temperatures. Maximally, 22.9 °C difference was observed when exposed concrete surface was covered with soil and shrubs at around 1400 hr. Among three plants, the temperature profile of the red plant was the highest, followed by turfing and the green plants. This indicates that LAI governs the amount of solar dissipation and soil surface temperatures. An interesting point to note is that the profile of the extensive control was the highest generally, followed by the profile of bare concrete and the intensive control (bare concrete surface does not have the highest profile). Since the three set-ups were exposed, the difference can only be attributed to the heat absorption ability of the various materials. Soil layer with no turf has the highest temperature because of the dark colour and a thin soil layer has less capability to absorb heat.

From Figure 8.107, it can be observed that the internal air temperatures measured in the boxes reached their peak values at around 1640 hr while the lowest values were found around 0745 hr. Temperature measured inside the box with exposed concrete surface was much higher from 1100 to 2400 hr and slightly lower from 0000 to 1100 hr compared with other boxes. The inverse situations, as expected, occurred on the rest of the experimental boxes with bare soil or planting. For the intensive system, the peak internal air temperature during the daytime was reduced by 7.5 °C when the exposed concrete surface was covered with bare soil and it was reduced by 9.3 °C when the green shrubs were introduced. It seems that 1.8 °C difference of internal air temperature was caused by plants. For the extensive system, the

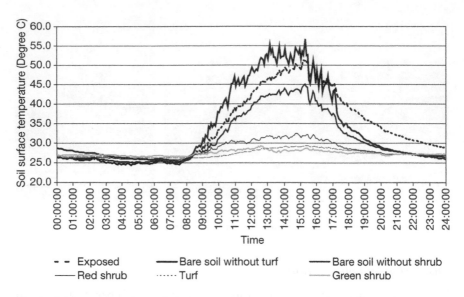

Figure 8.106 Comparison of surface temperatures measured at different locations
        (11 August 2003).

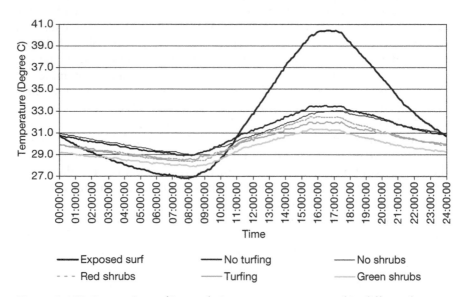

Figure 8.107 Comparison of internal air temperature measured in different boxes
        (11 August 2003).

peak differences were 7.1 °C and 8.1 °C when the bare soil and vegetation were introduced respectively. Maximally 1 °C difference of internal temperatures was caused by the turf during the daytime. At night, the internal temperature of the box with the exposed concrete surface was maximally 2.2 °C lower than the boxes with bare soil and maximally 1.6 °C lower than the planted boxes. This highlighted the effect of vegetation on stabilizing the fluctuation of internal air temperatures not only during the daytime but also during the night.

The correlation analysis was conducted among the environmental parameters (e.g. solar radiation, temperature, humidity, wind speed, etc.) derived from the weather station on site and soil surface temperatures of three types of plants with different LAIs. The soil surface temperatures are plotted against average solar radiation in Figure 8.108. It can be observed that the gradient of the trend line is the greatest for the red shrubs, followed by turfing and then the green shrubs. It seems that the soil surface temperature was quite sensitive to the changes in solar radiation for the red shrubs since they have small LAI value. For the turfing and the green shrubs, there is quite a large drop in gradients of the trend lines compared with that of the red shrubs. It can be inferred that the turfing and green shrubs were able to block off more solar radiation due to the substantial shading provided as reflected by their LAIs. It is also observed that the R^2 value decreases from the red shrubs ($R^2 = 0.72$) to the green shrubs ($R^2 = 0.52$). This means that in the case of the red shrubs, the link between soil surface temperature and

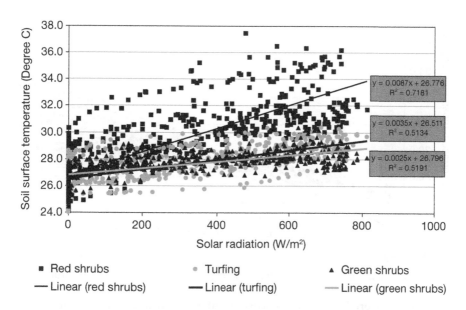

Figure 8.108 Correlation analysis between solar radiation and soil surface temperatures of three planted boxes (selected for seven clear days).

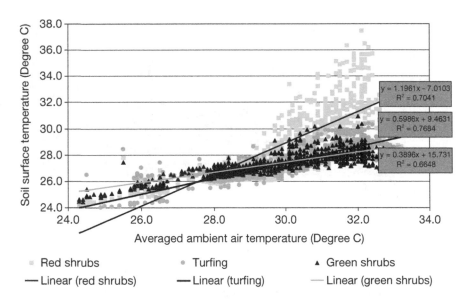

solar radiation explains 72 per cent of the variation in soil surface temperature. The remaining 28 per cent is variation in soil surface temperature when solar radiation is fixed. The correlationship is quite strong for the red shrubs. This shows that soil surface temperature varies more with solar radiation when LAI is low.

The soil surface temperatures of the three planted boxes are plotted against average ambient air temperature in Figure 8.109. Generally, it was observed that the red shrubs were most sensitive to the change of ambient temperatures, followed by turfing and green shrubs. When the ambient temperature was below 27.8 °C, the green shrubs had the highest soil surface temperature, followed by red shrubs and the turfing set-up. It seemed that the green shrubs had greater ability to stabilize the bounded air temperature. Therefore, they could not respond quickly to the cooler ambient air. Such conditions mostly occurred during early mornings and at night. When ambient temperature ranged from 28.2 °C to 29.9 °C, the red shrubs had the highest soil surface temperature. This indicates that red shrubs with lower LAI were susceptible to the ambient temperature. In the third segment when the ambient temperature was above 29.9 °C, the soil surface temperature of turfing also exceeded that of green shrubs.

The soil surface temperatures of the three planted boxes were plotted against average wind speed in Figure 8.110. Due to relatively low R^2 values, it can be deduced that wind speed in this case did not play a major role in

Figure 8.109 Correlation analysis between ambient air temperature and soil surface temperatures of three planted boxes (selected on seven clear days).

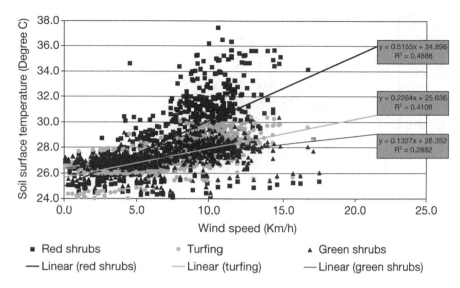

Figure 8.110 Correlation analysis between wind speed and soil surface
temperatures of three planted boxes (selected for seven clear days).

influencing the soil surface temperature. A possible reason may be that the
effect of solar radiation was too strong and overrode the effect of wind speed.
Figure 8.111 shows that soil surface temperatures of all three planted boxes
have a negative correlationship with the relative humidity measured through
the weather station on site. Generally, soil surface temperature dropped as
relative humidity increased.

During the second phase when the experimental boxes were cooled by an
air conditioner, relatively constant internal temperatures were successfully
created inside two boxes. The internal temperatures were maintained below
20 °C with a small fluctuation for 24 hours. This was to ensure that the
heat transfer was from outside to inside all the time. The surface temperatures
measured on the soil surface were not significantly influenced by the interior
condition due to reverse heat flow but rather the outdoor weather condition and the vegetation. The comparison of soil surface temperatures with
and without plants is presented in Figure 8.112. Maximally 10.9 °C
difference was observed at around 1440 hr. At night, about 0.9 to 1.2 °C
difference was obtained. These observations accord with the first phase of
measurement and indicate that the plants have a positive effect on dissipating incoming solar radiation during the daytime but delay the escape of heat
at night.

To further compare the cooling energy consumptions of different boxes,
an accumulative comparison of energy use was conducted (see Figure 8.113).
The result shows that during the early part of the day, the box with plants

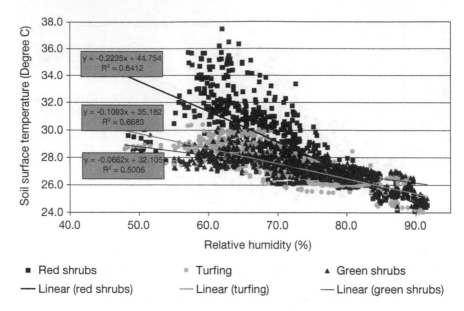

Figure 8.111 Correlation analysis between relative humidity and soil surface temperatures of three planted boxes (selected for seven clear days).

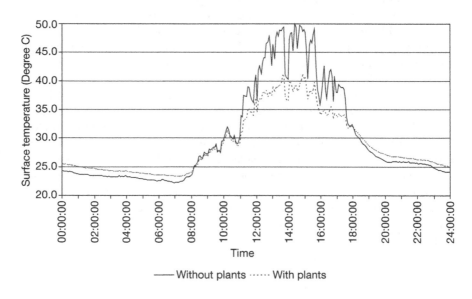

Figure 8.112 Comparison of soil surface temperatures measured with and without the red shrubs.

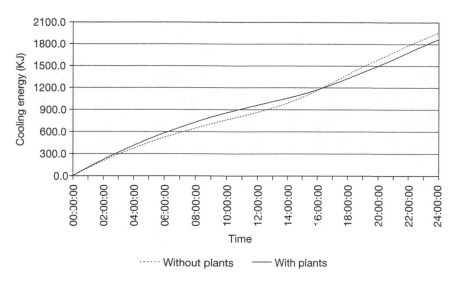

Figure 8.113 Comparison of cooling energy used for different boxes.

consumed more cooling energy than the box without plants. The thermal benefit of plants became more evident after 1630 hr when less cooling energy use was observed in the box with plants. Finally, 89 KJ cooling energy can be saved for a volume of 1 m³ of air, which is equal to 4.6 per cent of overall energy savings. Bear in mind, the box with plants was still covered with a soil layer then. The 4.6 per cent energy saving is purely from the sparse red plants with LAI of 1.5. It is expected that a higher contribution of every saving can be achieved by dense plants.

Experiment on the ground

Introduction

An open space in a nursery belonging to the National Parks Board (NParks) was chosen to carry out the experiment on the ground. It is paved with concrete and the area is about 250 square metres. The place is surrounded by shrubs and trees. Since it is located at a relatively higher level, there is no blockage of wind and no shading effect caused by landscape. Two set-ups, a horizontal one and a vertical one, were proposed. The experiment was carried out by monitoring the climatic parameters, bound air temperatures (within the foliage), soil surface temperatures and leaf surface temperatures on site (see Figure 8.114). The primary objective of the experiment was to work out usable regression models from the designed experimental set-ups according to plants' LAI values.

Figure 8.114 The experiment was carried out by monitoring many related
    parameters.

The experiment was carried out from 9 July to 18 November 2005 over a
period of around four months. There were two main phases in the
experiment. The first phase included two periods from 9 July to 1 September
and from 3 to 18 November. During this phase, the experiment was set up
mainly for 'warming up' all the equipment and for validation purposes. From
2 September to 2 November, all plants employed in the first phase were
replaced by new ones. Data collected during this phase were mainly used for
generating the regression models.

The horizontal set-up consisted of selected potted plants and a measuring
post mounted with seven temperature data loggers from just above the soil
surface to around 1400 mm with an interval of 200 mm. The potted plants
were placed close to each other with a dimension of 2 m (width) by 2 m
(length). The aim was to achieve expected LAI values. The measuring post
was placed in the centre of the foliage. Corresponding leaves located at
different heights (basically, the foliage was divided into low, medium and
high portions) were also measured. The density, or more accurately the LAI
values, was varied through the arrangement of the potted plants. With the
help of the LAI analyser, three arrangements of LAI values, which were 1, 3
and 5, were measured. The soil surface temperature was also measured in
the experiment.

The vertical set-up was made up of a wooden board, rows of plants and related instruments. The long axis of the partition board was orientated to face the west and the east orientations which are the extreme vertical cases under the local weather conditions. The dimension of the partition board was 1500 mm (length) by 1000 mm (height). Potted plants were placed closely in front of the board. The density (LAIs) of plants was varied by changing the rows of plants. In the experiment, single-row, double-row and triple-row arrangements were tested. The LAI values of the above arrangements were roughly 1, 3 and 5 respectively. It is worth mentioning that the LAI measurements conducted for the vertical set-up were done by tilting the lens of the LAI analyser towards the east or west instead of the sky. Bound air temperature was measured behind the plants at the two sides. The wooden board, which was around 50 mm thick, acted as a good thermal insulator and prevented the two sides from possible bound air temperature interference.

The weather station was set up near the two experimental set-ups but beyond their possible influential area. Similarly, the vertical and the horizontal set-ups were separated with enough distance to lower any possible mutual influence.

Discussion and observations

1. Horizontal set-up

The thermal performance of plants is very much governed by the density of foliage (LAI values). It has been fully tested and proven at both macro and micro levels by the previous background studies. In the experiment, the air temperatures within the foliages of various plants were again examined according to their corresponding LAI values. Figure 8.115 to Figure 8.117 illustrate the comparisons of temperatures obtained from open space (weather station) and within the plants over some clear days. Some general observations can be derived from the comparisons:

- The temperatures obtained at both the open space and within the plants are quite similar at night. It seems that the plants cannot vary bound air temperature within the foliage by a large amount at night. This observation accords with the long-term ones in which the lower whisker of the box-and-whisker plot indicates that air temperatures are not greatly reduced by the plants.
- The differences in temperatures between those measured at the open space and within the plants are mainly observed during the daytime, especially when the solar radiation is at its peak. This indicates that the plants, especially the dense ones, can provide effective sun shading and maintain a lower temperature condition within the foliages.
- The temperature difference varies from plant to plant according to their LAI values. The maximum difference observed in the sparse plants

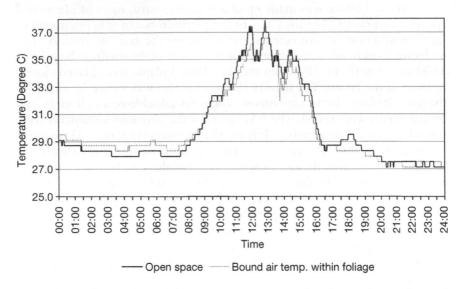

Figure 8.115 Comparison of the temperatures measured at the weather station and within the plant (LAI = 1) on a clear day.

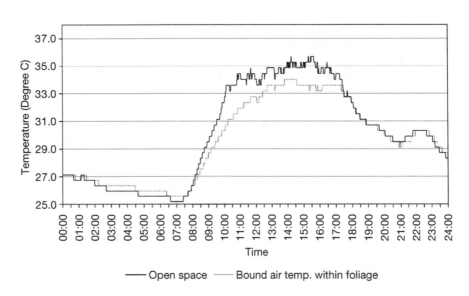

Figure 8.116 Comparison of the temperatures measured at the weather station and within the plant (LAI = 3) on a clear day.

(LAI = 1) is only 1.22 °C while it can be up to 4.58 °C within the dense ones (LAI = 5).

Figure 8.115 shows the comparison of temperatures obtained from the weather station and within the very sparse plants. It seems that the temperature difference between the sparse plants and the open space can only be experienced after around 1000 hr. But it can be observed as early as 0800 hr in the denser plants (see Figure 8.116 and Figure 8.117). A possible reason is that the altitude of the sun is relatively low in the morning and the sparse foliage cannot effectively intercept it during this period.

On the other hand, the bound air temperature within the sparse plants was also very sensitive to the fluctuation of the ambient air temperature while it was not easily observed in the denser plants. On the contrary, a time lag between the peak ambient temperature and the peak bound air temperature within the densest plants can be observed (see Figure 8.116). It is believed that the plants with high LAI values have the ability to maintain a relatively stable bound air condition within the foliage due to their outstanding ability for sun shading as well as evaporative cooling. The weak point of sparse plants is the lack of leaves. The incoming solar radiation cannot really be filtered, especially when the majority of leaf angle distribution is parallel to the sunlight.

Apart from measuring the bound air temperatures within the foliage, the surface temperatures of the leaves were also measured at the different plants. As mentioned earlier, the surface temperatures of the leaves was measured

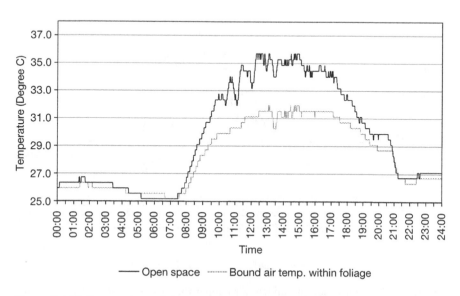

Figure 8.117 Comparison of the temperatures measured at the weather station and within the plant (LAI = 5) on a clear day.

not only at one measurement point but at five for the horizontal set-up. Ultimately, the leaf surface temperature is represented by the average value of the five measurement points. The comparisons of the ambient air temperatures obtained at the weather station, the bound air temperatures measured within plants and the average leaf temperatures are shown from Figure 8.118 to Figure 8.120.

It is clear that the average leaf surface temperatures measured at the different plants (LAI ranging from 1 to 5) are all lower than the corresponding bound air temperatures measured within the plants. This indicates that plants are actually 'cooling sources' which can consume external energy through the processes of photosynthesis and evapo-transpiration and bring tangible thermal benefits not only to the shaded area but also to the surroundings through long-wave heat exchange.

On the other hand, the difference between the average leaf surface temperatures and the bound air temperatures within plants is also closely related to the LAI values. With relatively lower bound air temperature during the daytime, the difference between the leaf surface temperatures and the bound air temperatures within the densest plants is not very apparent (see Figure 8.120) as compared to those obtained from the other two plants.

A correlation analysis between the bound air temperature within the foliage and other climatic parameters was done. It was found that the correlation between the bound air temperature within the foliage and wind speed was fairly weak. However, the solar radiation and ambient air

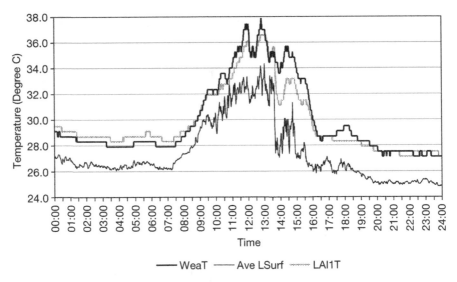

Figure 8.118 Comparison of ambient air temperature (weather station), bound air temperature and average leaf surface temperature within plants (LAI = 1) on a clear day.

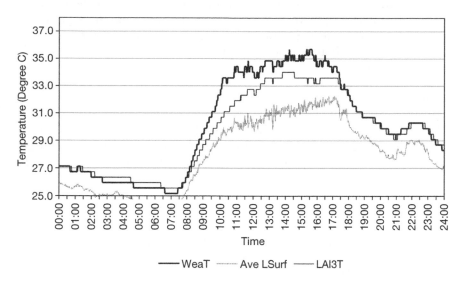

Figure 8.119 Comparison of ambient air temperature (weather station), bound air temperature and average leaf surface temperature within plants (LAI = 3) on a clear day.

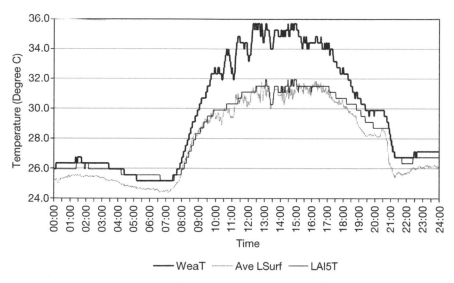

Figure 8.120 Comparison of ambient air temperature (weather station), bound air temperature and average leaf surface temperature within plants (LAI = 5) on a clear day.

temperature had a very strong correlation with the bound air temperature within the foliage. Based on the observations derived from the experiment, some regression models generated through Statistical Package for Social Sciences (SPSS) for both the morning and the afternoon sessions respectively were as follows:

Morning (bound air):

$Y_{lai1} = X_1 - [5.786LN(X_1) - 0.0002X_2 - 18.459]$ adjusted $r^2 = 0.698$ $\delta = 0.33$
$Y_{lai3} = X_1 - [12.286LN(X_1) - 0.002X_2 - 39.426]$ adjusted $r^2 = 0.933$ $\delta = 0.30$
$Y_{lai5} = X_1 - [19.56LN(X_1) - 0.001X_2 - 62.638]$ adjusted $r^2 = 0.982$ $\delta = 0.20$

where

Y_{lai1} = Bound air temperature within LAI 1 plants (°C)
Y_{lai3} = Bound air temperature within LAI 3 plants (°C)
Y_{lai5} = Bound air temperature within LAI 5 plants (°C)
X_1 = Ambient air temperature (°C)
X_2 = Solar radiation (w/m^2)
r^2 = Multi-correlation coefficient
δ = Standard error (°C)

Afternoon (bound air):

$Y_{lai1} = X_1 - EXP(-0.925X_1 - 0.004X_2 + 29.279)$ adjusted $r^2 = 0.234$ $\delta = 0.59$
$Y_{lai3} = X_1 - EXP(0.2X_1 + 0.001X_2 - 6.629)$ adjusted $r^2 = 0.942$ $\delta = 0.13$
$Y_{lai5} = X_1 - EXP(0.087X_1 + 0.001X_2 - 2.819)$ adjusted $r^2 = 0.974$ $\delta = 0.07$

where

Y_{lai1} = Bound air temperature within LAI 1 plants (°C)
Y_{lai3} = Bound air temperature within LAI 3 plants (°C)
Y_{lai5} = Bound air temperature within LAI 5 plants (°C)
X_1 = Ambient air temperature (°C)
X_2 = Solar radiation (w/m^2)
r^2 = Multi-correlation coefficient
δ = Standard error (°C)

   It is clear that the regression models for the sparse plants (LAI = 1) cannot be well established, especially during the afternoon session, as evident in the low adjusted R-square values. This further highlights the unstable bound air condition, which is not easily predicted, within the sparse foliage. On the other hand, very good multiple regression models have been built for the denser plants (LAI = 3 and 5). The adjusted r^2 can easily exceed 0.9. Similarly, the regression models related to estimation of leaf surface temperature in the plants was built up through SPSS for both the morning and the afternoon sessions respectively as follows:

Morning (leaf):

Y'_{lai1} = $1.487X_1 - 0.008X_2 - 15.642$ adjusted $r^2 = 0.728$ $\delta = 0.62$
Y'_{lai3} = $1.061X_1 - 0.005X_2 - 3.62$ adjusted $r^2 = 0.902$ $\delta = 0.46$
Y'_{lai5} = $1.039X_1 - 0.005X_2 - 3.202$ adjusted $r^2 = 0.902$ $\delta = 0.45$

where

Y'_{lai1} = Leaf surface temperature of LAI 1 plants (°C)
Y'_{lai3} = Leaf surface temperature of LAI 3 plants (°C)
Y'_{lai5} = Leaf surface temperature of LAI 5 plants (°C)
X_1 = Ambient air temperature (°C)
X_2 = Solar radiation (w/m²)
r^2 = Multi-correlation coefficient
δ = Standard error (°C)

Afternoon (leaf):

Y'_{lai1} = $3.76X_1 - 0.017X_2 - 89.344$ adjusted $r^2 = 0.412$ $\delta = 1.39$
Y'_{lai3} = $1.167X_1 - 0.005X_2 - 7.291$ adjusted $r^2 = 0.877$ $\delta = 0.39$
Y'_{lai5} = $1.339X_1 - 0.006X_2 - 12.651$ adjusted $r^2 = 0.888$ $\delta = 0.42$

where

Y'_{lai1} = Leaf surface temperature of LAI 1 plants (°C)
Y'_{lai3} = Leaf surface temperature of LAI 3 plants (°C)
Y'_{lai5} = Leaf surface temperature of LAI 5 plants (°C)
X_1 = Ambient air temperature (°C)
X_2 = Solar radiation (w/m²)
r^2 = Multi-correlation coefficient
δ = Standard error (°C)

The multiple regression functions of the leaf surface temperatures using ambient air temperature and solar radiation as the independent variables are quite similar to those of the bound air temperatures in the different plants. The multi-correlation is weak for the sparse plants (LAI = 1) while they are fairly good for the denser plants (LAI = 3 and 5).

2. Vertical set-up

The vertical set-up was also explored in order to establish usable regression models for vertically placed plants around buildings according to the plants' LAI values. Due to the constraints of the experiment, only the two 'worst case scenarios', the east and the west orientations which are vulnerable to extreme insolation, were measured. Actually, introducing plants into the two orientations is also applicable in reality. In order to avoid excessive solar heat gain in the tropical climate, buildings are normally designed with their long axis facing north–south. The eastern and the western façades are normally end walls with limited openings. Introducing plants into these end walls not only reduces the excessive heat gain but also avoids possible blockage of ventilation through openings.

Figure 8.121 shows the comparison between the ambient air temperatures measured from the weather station and the bound air temperatures behind the very sparse plants (LAI = 1) at the two orientations. Amazingly, the temperature measured behind the eastern greenery can be up to 4.2 °C higher than the ambient air temperature obtained from the weather station during the peak time (around 1000 to 1030 hr). After 1230 hr, the temperatures behind the eastern foliage began to drop. But only from 1500 hr to sunset was it significantly lower than the ambient air temperature. On the other hand, the maximum difference between the temperatures measured behind the western foliages and the weather station was around 3 °C and was observed at around 1630 to 1700 hr. Before around 1420 hr, the temperatures measured behind the western foliage were always lower than the ambient ones. As discussed previously, this can be explained by the combined effects of plants and orientation. The results can provide some hints for introducing plants, especially the sparse ones, into building façades in the

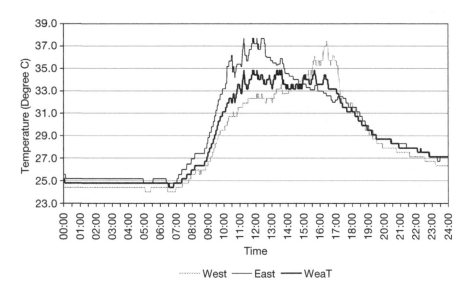

Figure 8.121 Comparison of the temperatures measured at the weather station and the bound air temperatures measured behind the plants (LAI = 1) at the western and the eastern orientations respectively on a clear day.

future. To reduce excessive solar radiation and create a low temperature condition near façades, vertical planting should be dense enough to intercept the low-altitude radiation for the eastern orientation. However, sparse plants can still contribute to façades facing west in the morning. They will take effect only during later afternoon when the low-altitude sunshine is experienced at the western orientation. Coupled with the long-term observations, it is believed that sparse plants can 'perform' better at the western orientation in creating a low temperature condition near façades.

The comparison of the temperatures measured behind the denser plants (LAI = 3) and the weather station is shown in Figure 8.122. With the increase of the LAI value, it is obvious that the thermal conditions at both the eastern and the western orientations are improved. The performance of plants on the eastern side is outstanding in mitigating the impact of orientation. Similar to the temperature profile made by the sparse plants at the same orientation, the temperature behind the eastern greenery increases immediately and reaches a peak of around 3.72 °C during the early morning. However, the temperature becomes lower than the ambient air temperatures after 1200 hr in the afternoon. The temperature measured behind the eastern vegetation can be 2.11 °C lower than the ambient temperature, which is observed at 1500 to 1530 hr in the afternoon. Moreover, due to the good protection of the plants at the eastern orientation, the peak temperature obtained at this orientation is similar to that obtained at the western orientation. The denser plants blur the effect of orientation, in this case at the eastern orientation. On the other

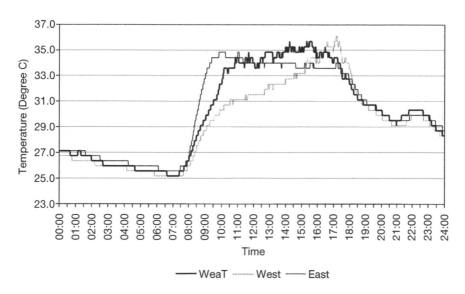

Figure 8.122 Comparison of the temperatures measured at the weather station and the bound air temperatures measured respectively behind the plants (LAI = 3) at the western and the eastern orientations on a clear day.

hand, better a cooling impact (up to 3.33 °C at around 1100 hr) from morning to around 1430 hr is also observed at the western orientation compared to the performance of the sparse plants at the same orientation.

Figure 8.123 shows the comparison of the temperatures measured at the weather station and behind the vertically placed vegetation (LAI = 5) at the two orientations. It can be concluded that the thermal conditions at the two orientations have been further improved. The improvement is significant enough to negate the orientation effect. The similar temperature profiles for the two orientations throughout the day can very well prove the preceding statement. There is no peak temperature observed at either early morning or late afternoon. On the contrary, the temperatures measured at the two orientations behind the plants show extreme similarity. The maximum cooling impact of up to 4.5 °C between the shaded bound air condition and the ambient air temperature is observed at both the eastern and the western orientations during the period from 1200 to 1400 hr when the peak solar radiation is experienced.

The comparisons among the ambient air temperatures, the temperatures measured behind the different plants and the average leaf surface temperatures obtained from the respective plants are presented from Figure 8.124 to Figure 8.129. Some general observations are:

- The leaf surface temperatures measured from the different plants are basically lower than the corresponding bound air temperatures measured

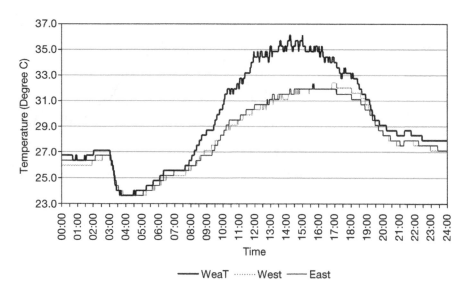

Figure 8.123 Comparison of the temperatures measured at the weather station and the bound air temperatures measured behind the plants (LAI = 5) at the western and the eastern orientations respectively on a clear day.

behind the plants except for those measured on the very sparse (LAI = 1) plants during the peak hours (early morning or late afternoon). Similar to the conclusion generated for the horizontal set-up, this indicates that the leaves of the plants work as cooling sources and thermally benefit the surrounding environment. Coupled with the shading effect, low bound air temperature conditions can be formed behind the plants at the two orientations.

- Obvious fluctuation of the leaf surface temperatures can be observed during the daytime while they are stable at night. This indicates that the surface temperatures of the leaves are sensitive to the solar radiation (the only measured climatic parameter which cannot be experienced at night) during the daytime.
- The night-time difference is bigger than the daytime one.
- The profiles of the leaf surface temperatures generated from the densest plants (LAI = 5) at the two orientations are quite similar to those obtained from the horizontal set-up with the similar LAI value. This indicates that the performances of the dense plants (high LAI values) can be very similar regardless of where they are placed.

Figure 8.124 and Figure 8.125 show the comparisons at the two orientations for the very sparse plants set-up. Compared to the bound air temperatures measured behind the plants, leaf surface temperatures are normally higher during the peak hours (such as early morning and later afternoon). This is slightly different from the profiles generated from the horizontal set-up with the sparse plants. But this does not mean that the leaves no longer act as cooling sources by then. It is the strong radiation which easily blurs the surface temperature readings. Compared to the same plants on the horizontal set-up, the sparse plants are weak in terms of intercepting the strong solar radiation at particular times since:

- the thin vertical layers of the leaves cannot shade each other effectively
- the vertically placed leaves are vulnerable to the low-altitude radiation.

With more leaves, the denser plants (LAI =3 and 5) can protect not only the hard surfaces behind them but also their own leaves. Therefore, lower leaf surface temperatures are observed all the time at the two orientations (see Figure 8.126 to Figure 8.129).

From the experiment results, it was found that the bound air conditions behind the densest plants (LAI = 5) at the vertical set-up were very stable while those behind the sparser plants (LAI = 1 and 3) were inconsistent at the two orientations. In generating reliable regression models, the unstable conditions should be excluded, simply because:

- Besides the measured meteorological data, many other unknowns also contribute to the unstable conditions. Since low-altitude solar radiation

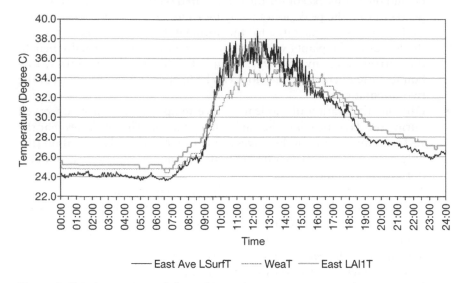

Figure 8.124 Comparison of the ambient air temperatures (weather station), the bound air temperatures and the average leaf surface temperatures within plants (LAI = 1) at the eastern orientation on a clear day.

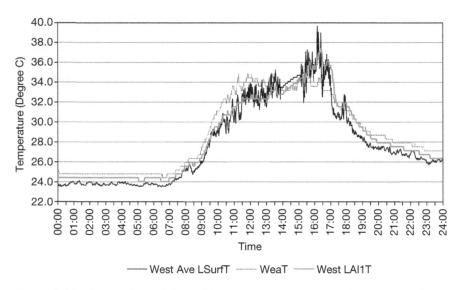

Figure 8.125 Comparison of the ambient air temperatures (weather station), the bound air temperatures and the average leaf surface temperatures within plants (LAI = 1) at the western orientation on a clear day.

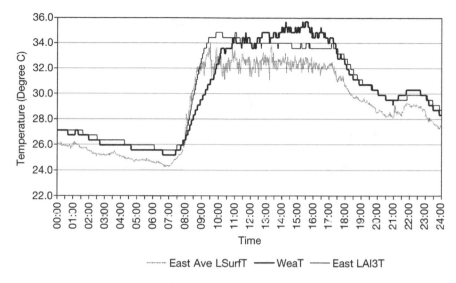

Figure 8.126 Comparison of the ambient air temperatures (weather station), the bound air temperatures and the average leaf surface temperatures within plants (LAI = 3) at the eastern orientation on a clear day.

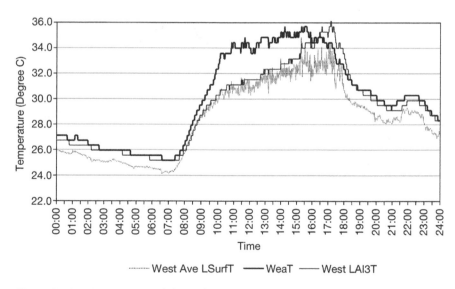

Figure 8.127 Comparison of the ambient air temperatures (weather station), the bound air temperatures and the average leaf surface temperatures within plants (LAI = 3) at the western orientation on a clear day.

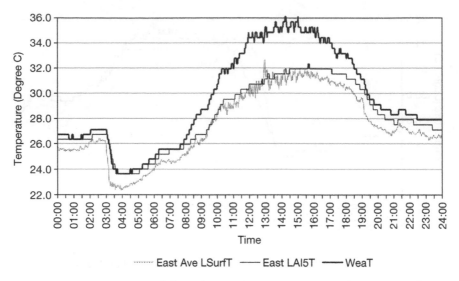

Figure 8.128 Comparison of the ambient air temperatures (weather station), the bound air temperatures and the average leaf surface temperatures within plants (LAI = 5) at the eastern orientation on a clear day.

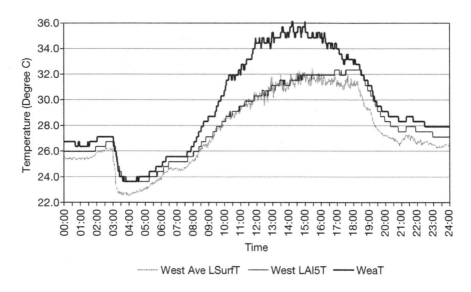

Figure 8.129 Comparison of the ambient air temperatures (weather station), the bound air temperatures and the average leaf surface temperatures within plants (LAI = 5) at the western orientation on a clear day.

can penetrate the vertically placed foliage and reach the vertical surface behind, the reflectivity, absorptivity and surface temperatures of the material behind the foliage should be taken into consideration. However, in the green sol-air temperature concept, predicting the bound air temperature within plants relies on the meteorological data.

• Defining the stable bound air condition is more important than predicting the unstable ones. The stable bound air condition is the benchmark which indicates that the vertical surface behind plants is becoming inconsequential. This means that the properties of the material behind plants will have a very limited impact on the variation of the bound air temperature. The benchmark can be applied to any vertical surface without considering the properties of building materials in each case. On the contrary, the unstable ones are more localized and cannot be easily applied to different conditions.

Therefore, regression models for vertically placed plants focused on the densest plants (LAI = 5) which can create a stable bound air environment between the plants and the hard surface in the experiment. Multiple correlation analyses between the temperature differences (ambient temperature measured at the weather station minus bound air temperature measured within foliage) and the weather variables which have been generated through SPSS for both the morning and the afternoon sessions respectively are generated as follows:

Morning (bound air):

$$Y_e = 0.27X_1 + 0.004X_2 + 17.922 \quad \text{adjusted } r^2 = 0.933 \; \delta = 0.27$$
$$Y_w = 0.175X_1 + 0.005X_2 + 20.444 \quad \text{adjusted } r^2 = 0.978 \; \delta = 0.16$$

where

Y_e = Bound air temperature behind LAI 5 plants at the eastern orientation (°C)
Y_w = Bound air temperature behind LAI 5 plants at the western orientation (°C)
X_1 = Ambient air temperature (°C)
X_2 = Solar radiation (w/m^2)
r^2 = Multi-correlation coefficient
δ = Standard error (°C)

Afternoon (bound air):

$$Y_e = X_1 - EXP(0.282X_1 + 0.0003X_2 - 8.903) \quad \text{adjusted } r^2 = 0.972$$
$$\delta = 0.09$$
$$Y_w = X_1 - EXP(0.312X_1 + 0.001X_2 - 10.655) \quad \text{adjusted } r^2 = 0.939$$
$$\delta = 0.20$$

where

Y_e = Bound air temperature behind LAI 5 plants at the eastern
orientation (°C)
Y_w = Bound air temperature behind LAI 5 plants at the western
orientation (°C)
X_1 = Ambient air temperature (°C)
X_2 = Solar radiation (w/m²)
r^2 = Multi-correlation coefficient
δ = Standard error (°C)

Similarly, regression models were generated for predicting the leaf surface
temperatures at the two orientations through SPSS as follows:
Morning (leaf):

Y'_e = $0.255X_1 + 0.005X_3 + 17.696$ adjusted $r^2 = 0.853$ $\delta = 0.36$
Y'_w = $0.132X_1 + 0.006X_2 + 21.064$ adjusted $r^2 = 0.938$ $\delta = 0.24$

where

Y'_e = Leaf surface temperature of LAI 5 plants at the eastern
orientation (°C)
Y'_w = Leaf surface temperature of LAI 5 plants at the western
orientation (°C)
X_1 = Ambient air temperature (°C)
X_2 = Solar radiation (w/m²)
r^2 = Multi-correlation coefficient
δ = Standard error (°C)

Afternoon (leaf):

Y'_e = $X_1 - EXP(0.154X_1 + 0.0002X_2 - 4.191)$ adjusted $r^2 = 0.952$
$\delta = 0.07$
Y'_w = $X_1 - EXP(0.166X_1 + 0.0002X_2 - 4.648)$ adjusted $r^2 = 0.952$
$\delta = 0.07$

where

Y'_e = Leaf surface temperature of LAI 5 plants at the eastern
orientation (°C)
Y'_w = Leaf surface temperature of LAI 5 plants at the western
orientation (°C)
X_1 = Ambient air temperature (°C)
X_2 = Solar radiation (w/m²)
r^2 = Multi-correlation coefficient
δ = Standard error (°C)

The above regression models obtained good adjusted r^2, which ranged from 0.853 to 0.978. The multi-regression functions can be applied to estimate leaf surface temperatures of plants at the two orientations within the standard error range of 0.07 to 0.36 °C.

Conclusion

The control experiments are an effective approach to play with some critical parameters found in the field measurements. Some significant findings derived from the two experiments are summarized as follows:

- As for the rooftop experiment, it was found that the impacts of plants, especially those with larger LAIs, were remarkable. The soil surface temperatures were reduced by 20.0 °C to 30.0 °C during the daytime. Plants help to reduce a huge amount of heat flux entering the building structure. Cooling energy can be saved. Although only 4.6 per cent of overall energy savings was observed on boxes planted with small LAI vegetation, it is expected that more energy can be saved when dense plants (large LAI) are applied. The interactions among weather parameters on soil surface temperatures are complicated since the effect of some parameters can be overridden. However, some of the surface temperatures under the vegetation have relatively strong correlation with solar radiation and ambient air temperature.
- As for the horizontal set-up in the experiment carried out on the ground, it was found that the LAI values also govern the performances of the different plants. The maximum difference observed in the sparse plants (LAI = 1) was only 1.22 °C while it was up to 4.58 °C within the dense ones (LAI = 5). The cross-comparisons between the different plants also follows the sequence of their corresponding LAI values. It seems that a stable bound air temperature condition can be formed within the denser foliages (LAI = 3 and 5). But it fails to be formed with the very sparse plants (LAI = 1). The leaf surface temperatures are mostly lower than the corresponding bound air temperatures within the foliage. This indicates that the leaves are cooling sources during the daytime. Coupled with their outstanding shading effect, the low-temperature leaves are the reason for the lower bound air temperature.
- As for the vertical set-up in the experiment carried out on the ground, it was found that the thermal impacts of plants over the vertical set-up were slightly different from those over the horizontal set-up. A stable bound air condition is not easily achieved behind the plants unless the LAI value reaches a higher level compared to that observed in the horizontal set-up. In the experiment, it was found that the threshold of LAI value for achieving a stable bound air condition was 5. The low-altitude solar radiation that occurs during early morning or late afternoon is not easily intercepted by the plants with horizontally

distributed leaves even through the LAI values can be up to 3. On the other hand, interception of diffused radiation will be easier compared to stopping direct radiation by the sparse foliage. The observation is usable and can be applied to other orientations, such as the north and the south, where a longer period of only diffused radiation within the year is experienced. At these orientations, introducing relatively sparse plants is still reasonable since the plants can effectively intercept diffused radiation and create a better thermal environment near façades. Compared to the bound air temperatures measured behind the plants, leaf surface temperatures are normally higher during the peak hours (such as early morning and late afternoon) for the sparse plants (LAI = 1). This does not mean that the leaves stopped acting as cooling sources then. It is the strong radiation which easily blurs the surface temperature readings. With more leaves, the denser plants (LAI = 3 and 5) can protect not only the hard surfaces behind them but also their own leaves. Therefore, lower leaf surface temperatures are observed all the time at the two orientations.

Index

Pages with figures are designated with the number followed by *fig*.

adaptive behaviour 11–12
airflow 19, 19*fig*, 60, 87
Akbari, H. 74
albedo 59
American Society of Heating, Refrigerating and Air conditioning Engineers 7, 12
anthropogenic heat source 54, 133, 161–2

balcony gardens 42*fig*
Bamboo house 21, 22*fig*
Barges, Hermann 77
Barkman, J. J. 82
Bay, Joo Hwa 17–18, 24–5
Bedford scale 12
Bernatsky, A. 88
biodiversity 68
bound air condition 251–2
boundary layer heat island (UBL) 56
Bowen ratio 85
Bridgman, H. 53
buildings: vernacular tropical 10, 21–3, 22*fig*; sick 11, 30; tropical 27, 28–9*figs*, 30, 30–1*figs*, 32; and solar radiation 32, 59, 92; and climate 51, 52*fig*, 53–4, 55*fig*; and plants 96–103, 99–101*figs*; and vertical landscaping 218; and green experiments 244
Bukit Batok Nature Park, Singapore (BBNP) 132–3, 132*t*, 133–8*figs*, 135
Bukit Timah reserve, Singapore 38, 39*fig*, 45, 48

Ca, V. T. 89
canopy layer heat island (UCL) 56

canyon geometry 54, 60
CBD area, Singapore 110, 112, 122, 126, 128, 130–1
Changi airport, Singapore 110, 112–13, 117*fig*, 118, 119–20*figs*, 122, 130–1
Changi Business Park (CBP) 150, 153–5, 154*fig*, 156–9*figs*
Chen, Y. 92
Chia, L. S. 56
Chinatown, Singapore 24*fig*
chlorosis 71
city parks 38
Clementi Woods Park, Singapore (CWP) 132, 132*t*, 133*fig*, 135*fig*, 138–9, 139–40*fig*, 141–2, 143–4*figs*, 145–6, 146*fig*
climate: introduction 3; tropical 3, 4*fig*; and buildings 51, 52*fig*, 53, 55*fig*; responsive design 61, 63, 63–4*figs*; and plants 82–3, 85–93; buildings model 96–103, 99–101*figs*
coefficient of performance (COP) 86
Concept Plan 2001 148
Crassulacean Acid Metabolic (CAM) 193

deforestation 35–6, 37*fig*, 46*fig*, *see also* trees
Delta T (surface temperature difference) 218, 219*fig*
Dodson, J. 53
Dwyer, J. F. 76

emissivity 59
Emmanuel, M. R. 20, 32, 63
energy consumption: cooling 19–20, 19–20*figs*; and shading 30; buildings

53; and UHI 58; and greenery 74; and parks 147; savings 161, 163–4, 163*fig*
Envi-met 142, 147, 154
environment 69–71, 70*fig*, 99*fig*, *see also* climate
evapo-transpiration 89
evaporative cooling source 54
Eyring, C. F. 73

factory buildings 218, *see also* buildings
Fanger, P. O. 9
fossil fuels 53, 58
Frankfurt 88

garden cities 102–3
Garden City Action Committee, Singapore 47
Ginza Plaza, Singapore 145
Givoni, B. 20, 75, 77, 86
global warming 83
Greece 91
Green City movement 103
greenhouse effect 54, 58
Greenbelt concept 41
greenery *see* plants; trees; tropical plants
Greenway concept 41
Gupta, A. 17

hanging-down plants 41, 43*fig*, 44*t*, *see also* vertical landscaping
Harazono, Y. 90
heat flux 187–8, 218, 219*fig*, 225
Hoffman, M. E. 88
Holm, D. 92
Housing and Development Board, Singapore (HDB) 26, 27–8*figs*, 132–3, 135, 139, 139*fig*, 142, 146
Howard, Ebenezer 102
Howard, Luck 53, 56
Hoyano, A. 92
Hyde, R. 61

India 91
industrial revolution 102
International Business Park, Singapore (IBP) 169, 170*fig*, 171–5, 172–6*figs*, 177, 178*figs*, 179

Japan 75, 88–9
Jauregui, E. 88
Johnston, J. 76

Kalae house 21, 22*fig*
Kallang Airport, Singapore 24
Kampong house 21, 22*fig*
Kaushik, S. C. 91
Kawashima, S. 88–9
Kent Vale blocks, Singapore 139, 140*fig*, 145
Knepper, C. A. 74
Koenigsberger, O. H. 83
Köpen climate classification 3
Kumar, R. 91

Landsat-7 satellite 109, 110–11*figs*
Landsberg, H. E. 56
landscaping 39–41, 42–4*figs*, 45, 75, 205, *see also* vertical landscaping
Leaf Area Index (LAI): explanation 85, 91; urban parks 132, 139–40, 140–1*figs*, 146–7; intensive rooftop gardens 171; metal roofs 196–7; vertical landscaping 205; green experiments 235–7, 240–1*figs*, 242–7, 244*fig*, 246*fig*, 248–50*figs*, 251–4
Lee Kuan Yew 47
Letchworth, United Kingdom 102

McGregor, G. R. 36
McPherson, E. G. 74
Macritchie reservoir, Singapore 122
macroclimate 82
Master Plan 2003 148
mean radiation temperature (MRT) 172, 175, 179
mesoclimate 82
Mexico City 88
Minke, G. 72
'moisture island effect' 87
mulching 71

National Parks Board, Singapore (NParks) 48, 235
National University of Singapore, Building Department 227
necrosis 71
neighbourhood parks 39*fig*
Newton, J. 76
Niachou, A. 91
Nieuwolt, S. 36
non-atmospheric heat island 56
NTT Urban Development Corp. (Japan) 75

Oke, T. R. 56, 58

Olgyay's model 97–8, 98*fig*
Onmura, S. 91
overpasses 209, 209*fig*
oxygen 72
ozone 18, 71

Park and Recreation Department,
Singapore (PRD) 47
Parker, J. H. 74
Parsons, R. 76
Picot, X. 87
Planck's blackbody formula 109
plants: modular 41, 44*fig*,, 44*t*;
wall-climbing 41, 43*fig*, 44*t*; and
buildings 75–6, 99–103, 107; and
climate 82–3, 85–93; and solar
radiation 86–7, 205, 206–9*figs*, 209,
212–13, 215–16, 217*figs*; UHI
measurement 126, 128; urban
parks 142, 146; and trees 150, 155;
and intensive rooftop gardens
171–2, 175, 177, 179; and
extensive rooftop gardens 192–3,
196, 197–200*fig*, 203; vertical
landscaping 204–5, 206–10*figs*,
209, 212–13, 218, 220, 222–6,
222*fig*, 224*fig*; green experiments
227–9, 228*fig*, 230–4*figs*, 231–7,
238–41*figs*, 239–40, 242–7,
244–6*figs*, 248–50*figs*, 251–4
plants–climate–buildings model
101–2
pollution 18
PowerDOE energy simulation
programme 177
public green areas 88–90, 90*fig*
Punggol site: trees 148, 149*fig*, 150,
151–2*figs*; intensive rooftop gardens
165, 167–8, 167–9*figs*; extensive
rooftop gardens 180–1, 181–4*figs*,
185, 186–91*figs*, 187–9, 192–4,
192*fig*, 194–5*figs*, 196–200,
197–202*figs*, 202–3

radiation 53–4
radiative forcing agents (RFA) 18
Raffles, Sir Stamford 45
rainfall 35
rainforest distribution 84*fig*
relative humidity (RH) 4–6
Roaf, S. 21
roofs: rooftop gardens 23, 33, 41–2*figs*,
171, 192–3, 196, 197–200*figs*, 203;
green 71, 73–4, 76, 90–2, 91*t*; metal

196–200, 197–202*figs*, 202–3; and
solar radiation 200
Rusk, Dr Howard A. 76

Santamouris, M. 85
Selatar station, Singapore 112, 115*fig*
Sembawang station, Singapore 112–13,
114*fig*
Seng Kang site, Singapore 148, 149*fig*,
150, 151–2*figs*
shading: tropical buildings 27, 28–9*figs*,
30, 30–1*figs*, 32; plants and climate
89; extensive rooftop gardens 185;
vertical landscaping 205, 206–9*figs*,
209, 210*fig*, 218, 218*fig*; green
experiments 237, 247
'shadow umbrella' strategies
(Emmanuel) 32
Shashua-Bar, L. 88
shophouses 24, 24*fig*
sick buildings 11, 30, *see also*
buildings
Simpson, J. R. 74
Singapore: Clementi Woods Park *see*
Clementi Woods Park, Singapore
(CWP); energy consumption 20;
Chinatown 24*fig*; Kallang Airport
24; and modern tropical buildings
24–6, 25–6*figs*, 28–31*figs*;
shophouses 24, 24*fig*; Housing and
Development Board (HDB) 26,
27–8*figs*, 132–3, 135, 139, 139*fig*,
142, 146; Bukit Timah reserve 38,
39*fig*, 45, 48; green history 45,
46–7*figs*, 47; Garden City Action
Committee 47; Park and Recreation
Department (PRD) 47; garden city
48–9, 102; National Parks Board
(NParks) 48, 235; Urban
Redevelopment Authority (URA) 48,
148; Singapore Meteorological
Service (MSS) 56; UHI study 56;
attracting wildlife to 72; vertical
landscaping 92, 204–5, 204*fig*,
206–9*figs*, 209, 211–13, 214–19*figs*,
215–16, 218, 220, 221–5*figs*,
222–6; case study introduction 107;
urban/rural 108*fig*; satellite studies
109–12, 110–11*figs*; CBD area 110,
112, 122, 126, 128, 130–1; Changi
airport 110, 112–13, 117*fig*, 118,
119–20*figs*, 122, 130–1; historical
weather 112–13, 114–17*figs*, 118;
Selatar station 112, 115*fig*;

Sembawang station 112–13, 114*fig*;
Tengah station 112–13, 116*fig*, 118,
119–20*figs*, 130; temperature
mapping 118, 119–21*figs*, 121–2,
123–4*figs*, 125–8, 127*fig*, 130–1,
130*t*; Macritchie reservoir 122;
one-route survey (Singapore) 125*fig*;
four-route survey (Singapore) 126*fig*;
Tukey- Kramer test 128, 129*t*;
zoning 128, 128*fig/t*; Bukit Batok
Nature Park (BBNP) 132–3, 132*t*,
133*fig*, 134–8*figs*, 135; urban parks
132–3, 132*t*, 133*fig*, 135–6, 138–9,
141–2, 145–7; Kent Vale blocks
139, 140*fig*, 145; Ginza Plaza 145;
Seng Kang site 148, 149*fig*, 150,
151–2*figs*; trees 148, 150, 153*fig*,
154–5, 160–4; Changi Business Park
(CBP) 150, 153–5, 154*fig*,
156–9*figs*; intensive rooftop gardens
165, 166*fig*, 167–9, 171–5,
172–6*figs*, 177, 179; International
Business Park (IBP) 169, 170*fig*,
171–5, 172–6*figs*, 177, 178*figs*, 179;
extensive rooftop gardens 180–1,
181–4*figs*, 185, 186–91*figs*, 187–9,
192–4, 192*fig*, 194–5*figs*, 196–200,
197–202*figs*, 202–3; green
experiments 227–9, 228*fig*,
230–4*figs*, 231–7, 236*fig*,
238–41*figs*, 239–40, 242–7,
244–6*figs*, 248–50*figs*, 251–4;
National University, Building
Department 227; weather station
228, 231, 244–5, 245–6*figs*,
248–9*figs*
Singapore Meteorological Service (MSS)
56
Smith, M. 96
smog 58–9, 68, 71, 76
solar radiation: description 5–6; and
tropical buildings 32; climate and
buildings 51, 56, 59, 92; and plants
86–7, 205, 206–9*figs*, 209, 212–13,
215–16, 217*figs*; parks 141–2; and
roofs 200; and green experiments
231–3, 231*fig*
Statistical Package for Social Sciences
(SPSS) 242–3
Stein, Clarence 102
stilts 23
Stoutjesdijk, P. H. 82
Sumatra 35
surface UHI 58, 118

suspended particulate matter (SPM) 18

Taha, H. 74
Takakura, T. 90–1
Tama Central Park, Tokyo 89
Tay, K. S. 20
Tel Aviv, Israel 88
Tengah station, Singapore 112–13,
116*fig*, 118, 119–20*figs*, 130
thermal comfort 7–15, 10–11*figs*,
12–14*t*
Tongkonan house *see* Torajan house
topoclimate *see* mesoclimate
Torajan house 21, 22*fig*
Toronto Food Policy Council (TFPC)
75
trees 72–4, 148, 153–5, 161–4,
211–14, 211–13*figs*, 215–17*figs*,
226
tropical architecture 30, 32
tropical plants: climate 8*fig*; urban
importance 21, 23, 33, 59;
rainforest 35–7, 36, 36–7*figs*;
urban artificial 38–40, 38–44*figs*,
44*t*, urban natural 37–8, *see also*
plants
Tso, C. P. 56
Tuas Avenue, Singapore 155, 160–2,
160–2*figs*

Ulrich, R. S. 76
urban design strategies 32–3
urban heat island effect (UHI):
description 5; climate and buildings
53–6, 54–5*figs*, 57*fig*, 58–9, 63;
buildings and plants 69; satellite
studies of Singapore 109–12,
110–11*figs*; historical weather,
Singapore 112–13, 114–17*figs*, 118;
temperature mapping, Singapore
118, 119–21*figs*, 121–2, 123–4*figs*,
125–8, 127*fig*, 130–1; metal roofs,
Singapore 196–7
urban parks 132, 139–40, 140–1*figs*,
146–7
Urban Redevelopment Authority,
Singapore (URA) 48, 148
urbanization 17–21, 18*fig*, 23, 48,
68–78, 70*fig*

ventilation 55, 59–61, 62*fig*
vernacular tropical buildings 10, 21–3,
22*fig*, *see also* buildings
vertical landscaping: tropical plants

40–1, 44*t*; buildings 74; plants 92–3, 92*t*; in Singapore 204–5, 204*fig*, 206–9*figs*, 209, 211–13, 214–19*figs*, 215–16, 218, 220, 221–5*figs*, 222–6
Vitruvian tripartite model 97, 97*fig*
void decks 26, 27–8*figs*
volatile organic hydrocarbons (VOCs) 71
Voogt, J. A. 58

wall-climbing plants 41, 43*fig*, 44*t*, *see also* plants
Warner, R. 53
water 71–3
weather station, Singapore 228, 231, 244–5, 245–6*figs*, 248–9*figs*

Welwyn Garden City, United Kingdom 102
'white noise' 76
wildlife 72
Wilmers, F. 92
wind 5–6, 51, 55, 60, 87, 146
wind tunnel studies 60–1*figs*
windows 33
Witter, G. 72
World Health Organisation (WHO) 18
Wright, Henry 102

Yamada, I. 35
Yeang, Ken 40, 96

zoning, Singapore 128, 128*fig/t*